Cambridge Elements ≡

Elements in the Philosophy of Biology
edited by
Grant Ramsey
KU Leuven
Michael Ruse
Florida State University

REDUCTION AND MECHANISM

Alex Rosenberg
Duke University

CAMBRIDGE
UNIVERSITY PRESS

CAMBRIDGE
UNIVERSITY PRESS

University Printing House, Cambridge CB2 8BS, United Kingdom

One Liberty Plaza, 20th Floor, New York, NY 10006, USA

477 Williamstown Road, Port Melbourne, VIC 3207, Australia

314–321, 3rd Floor, Plot 3, Splendor Forum, Jasola District Centre,
New Delhi – 110025, India

79 Anson Road, #06–04/06, Singapore 079906

Cambridge University Press is part of the University of Cambridge.

It furthers the University's mission by disseminating knowledge in the pursuit of
education, learning, and research at the highest international levels of excellence.

www.cambridge.org
Information on this title: www.cambridge.org/9781108742313
DOI: 10.1017/9781108592949

First published 2020

A catalogue record for this publication is available from the British Library.

ISBN 978-1-108-74231-3 Paperback
ISSN 2515-1126 (online)
ISSN 2515-1118 (print)

Reduction and Mechanism

Elements in the Philosophy of Biology

DOI: 10.1017/9781108592949
First published online: June 2020

Alex Rosenberg
Duke University

Abstract: Reductionism is a widely endorsed methodology among biologists, a metaphysical theory advanced to vindicate the biologist's methodology, and an epistemic thesis those opposed to reductionism have been eager to refute. While the methodology has gone from strength to strength in its history of achievements, the metaphysical thesis grounding it remained controversial despite its significant changes over the last seventy-five years of the philosophy of science. Meanwhile, antireductionism about biology, and especially Darwinian natural selection, became orthodoxy in philosophy of mind, philosophy of science, and philosophy of biology.

This Element expounds the debate about reductionism in biology, from the work of the post-positivists to the end-of-the-century debates about supervenience, multiple realizability, and explanatory exclusion. It shows how the more widely accepted twenty-first-century doctrine of "mechanism" – reductionism with a human face – inherits both the strengths and the challenges of the view it has largely supplanted.

Keywords: reductionism, mechanism, special sciences

ISBNs: 9781108742313 (PB), 9781108592949 (OC)
ISSNs: 2515-1126 (online), 2515-1118 (print)

Contents

Introduction

Reductionism is a metaphysical thesis, a claim about explanations, and a research program.

The research program is pursued by scientists, who point to great achievements that encourage its pursuit while recognizing the important challenges it still faces. What this program demands is that biology and the life sciences generally employ an opportunistic, top-down research strategy, always seeking more fundamental explanations that will complete, correct, or otherwise improve on previous explanatory achievements. Reductionism is often treated as also requiring an equivalent research program in the behavioral and social sciences, which, following Fodor (1974), have come to be called the "special sciences." This extension is not unwarranted, since the special sciences are devoted to the study of one or a small number of biological kinds: *hominins* and higher primates.

Reductionism as a research program among biologists doesn't impose a bottom-up strategy that requires all biological research to begin at the molecular level. It only requires it to end there. The actual reductions these biologists have provided are illustrations to them and to reductionist philosophers of how biological explanation at its best proceeds. The successes motivate biologists far more than any philosophical arguments. Among these successes are at least the following: the molecular account of respiration all the way down to the oxygen affinity of the iron molecule in the four-unit adult and fetal hemoglobin protein molecules, their allosteric properties, and the roles of coenzymes such as 2,3-diphosphoglycerate (DPG) in their positive cooperativity (Perutz, 1962); the central dogma of molecular genetics (Watson, 1965) and its associated achievements in the explanation of heredity and somatic gene regulation; the specific role of somatic gene regulation in embryological development (Lawrence, 1992); sensory transduction in the visual system (Montell, 2003); and the less spectacular successes of the physical understanding of photosynthesis as a chemical process – the Calvin cycle – and the even earlier detailing of the mechanism of cellular metabolism in the tricarboxylic acid (TCA) or Krebs citric acid cycle. By itself, a list of successes of this sort is at best a relatively weak inductive argument for reductionism as the methodology of a research program that will succeed everywhere in the life sciences. A few biologists have tried to provide more arguments for why such a program should face no limits and be expected to succeed everywhere in the life sciences. Of these, the most visibly explicit advocates of reductionism were the Nobel laureates J. Monod (1971) and F. Crick (1966). By and large, their arguments were metaphysical: the nature of reality as being wholly physical motivated their reductionism. But metaphysics doesn't much concern biologists, when they even notice it at all. Philosophers, however, are not indifferent to it. Philosophers who endorse the research program on which biologists such as Monod and Crick placed their bet ground their confidence in its eventual metaphysical vindication, which they identify as physicalism: All of the facts empirical science deals with are physical facts – facts about physical matter and physical fields.

Reductionism's biological and philosophical opponents reject the research program as misguided because they insist there are many excellent irreducible explanations in biology. In the words of one of the most prominent of anti-reductionists, reductionism's "mistake consists in the loss of understanding through immersion in detail, with concomitant failure to represent generalities that are important to 'growth and form' [invoking D'Arcy Wentworth Thompson's expression]" (Kitcher, 1999, p. 206). As we will see, they proffer examples of such fully adequate but irreducible explanations.

Many (perhaps most) philosophers who reject reductionism as a research program or a thesis about explanation embrace a metaphysical commitment to physicalism. This is especially true among philosophers of mind and philosophers of psychology who do not wish to endorse dualism or the existence of mind or mental events, states, or processes distinct from physical events, states, or processes. Physicalist antireductionism is in fact the most widely held view in the philosophy of mind and philosophy of psychology, as well as a widely accepted view in biology, as we shall see.

But it is important to bear in mind that there are some philosophers of science, including a number of philosophers of biology, who reject reductionism for other reasons. Many of them call themselves pluralists largely because they reject the suggestion that there is a single epistemic standard or methodology underlying science, especially one that requires strong consilience or even logical compatibility between disciplines and their claims. These pluralists emphasize the corrigibility and tentativeness of all scientific findings, and they urge that serious consideration of a large number of different, competing, and conflicting theoretical and experimental results is healthy. Thus, they reject physicalism as the reflection of unacceptable claims for the hegemony of results and methods characteristic of the physical sciences. Pluralism is committed to (tentatively) accepting the knowledge claims of the apparently irreducible findings, theories, and explanations they identify in biology. The train of their argument proceeds from their acceptance of one or another theory, model, regularity, or explanation in biology as well established, together with their claim that it is irreducible, to the intermediate conclusion that as philosophers we should endorse them as true or well warranted owing to their widespread acceptance among biologists. The last step in these philosophers' argument is that the acceptance of these irreducible biological claims disconfirms physicalism since their irreducibility is incompatible with physicalism. These philosophers are metaphysical pluralists (see Dupres, 1993).

These philosophers of biology share with physicalists the suggestion of Quine (1960) that philosophy should be strongly guided by science, but they differ from him on which sciences should shape metaphysics and epistemology. Thus, pluralists in effect adopt a variety of epistemological standards that allow for the diverse range of mutually irreducible, competing, and conflicting biological achievements they endorse (see Kellert et al., 2006).

Most philosophers of science accept a more monistic epistemology, one that puts a premium on predictive success – in both range and precision – and that is more fully satisfied by physics than any other discipline. Accordingly, they are driven to attach the greatest epistemic weight to physics and to impose something like the epistemic standards it meets on other sciences. This makes them physicalists about the nature of reality and grounds their rejection of pluralism. For most philosophers, therefore, the reductionism/antireductionism debate is one that presupposes a shared agreement on physicalism. This will be the operative assumption of the remainder of this Element.

Antireductionists accept the physicalist metaphysics reductionists rely upon, but they hold it innocent of the implications for explanation in biology that the research program of reduction infers. In fact, as noted, physicalist antireductionism is the ruling view in the philosophy of biology and in its cognate (sub) disciplines of the philosophy of psychology, the philosophy of neuroscience, and the behavioral and social sciences – the special sciences – generally.

Both of the parties to the philosophical dispute have long grounded their arguments on their respective analyses of many features of biology as a domain and of the science of biology as its study. The debate between reductionists and antireductionists has therefore had important spillover effects, as it has raised many of the most fundamental questions in the philosophy of science: the nature of biology's distinctive concepts, the scope for biological laws, the structure of biological explanations, and the role of Darwin's theory of natural selection in organizing the discipline and its domain of inquiry.

The metaphysical thesis reductionists advance is that all biological facts are fixed by physical and chemical facts; there are no nonphysical events, states, or processes, and so biological events, states, and processes are "nothing but" physical ones. This is physicalism. If the physical facts "fix" the biological ones, then, claims the reductionist, physical science should be able to reductively explain all of the latter facts. The research program can be framed as a methodological prescription that follows from the claim about explanations: Seek explanations for higher-level biological phenomena in discoveries about lower-level ones. When stated as a claim about fact-fixing by physics, few antireductionists dispute reductionism's metaphysical claim, but all reject its demands on explanation and so its methodological moral for improving explanation. To a first approximation, what reductionists and antireductionists disagree about is whether explanations in *functional* biology can be or need to be deepened or improved by, completed, or perhaps replaced by explanations in terms of molecular biology. The issue thus raised is whether metaphysical reductionism (i.e., physicalism about the biological), a thesis all claim to accept, requires explanatory reduction and necessitates a reductionistic research program in the life sciences.

Functional biology is the study of phenomena under their functional kind descriptions, such as *organism, organ, tissue, cell, organelle*, and *gene*. Molecular biology is the study of certain classes of organic macromolecules that may be characterized by their chemical structure. This distinction is not entirely satisfactory, for many of the kinds identified in molecular biology are also individuated functionally. What makes a kind *functional* is that its instances are the products of an evolutionary etiology – a history of random variation and natural selection. Since natural selection operates at the macromolecular level, some (indeed, many) of molecular biology's distinctive kinds will be functional as well. This makes the functional/molecular distinction a misleading one for the purposes of debates about reduction in biology. Alas, the terminology has become established and no obvious improvement on it has emerged. We will employ the functional/molecular distinction hereafter on the understanding that much molecular biology attributes functions to macromolecules. Indeed, the fact that they have functions will be crucial to the most important issues to be broached in this Element.

As noted, disagreement between reductionists and their opponents over the adequacy of explanations in functional biology drives their methodological disagreement, with important consequences for the research program of biology. The reason for this is simple: If the aim of science is explanation and many explanations in functional biology are adequate, complete, and correct, then the methodological prescription that we *must* search for molecular completions or corrections of these functional explanations in molecular processes will be unwarranted. Consequently, molecular biology will not be the inevitable foundation for every compartment of functional biology. But if the aim of science is explanation and functional explanations are either incorrect or incomplete and molecular explanations are either (more) correct or (more) complete, then biology must act on the methodological prescription that we should seek macromolecular explanations of all functional biological phenomena. All biologists who seek complete and correct explanations will have eventually to be molecular biologists.

It is important to note here a difference between reductionism and eliminativism, a quite different thesis with which reductionists are often mistakenly saddled even as they wish to avoid it. In its strongest form, eliminativism is the indefensible thesis that all biology is molecular biology – that molecular biology provides not only the *explanans* (what does the explaining), but also uncovers all of the facts to be explained (the *explanantia*). Strong eliminativism thus affords no explanatory or descriptive role to functional biology. Reductionism allows for both. It only insists that higher-level or functional biological explanations be grounded in macromolecular ones.[1]

Most philosophers of science who take sides in the reductionism/antireductionism debate do so because they have undertaken a study of the biological domain and the discipline that seeks knowledge of it. As a result, they think one or the other of the two more adequately illuminates the domain and the discipline. But a few philosophers may embrace reductionism on the basis of a purely philosophical, metaphysical argument due to Jaegwon Kim (2005). The argument is quite simple. It seeks to show that, when packaged together, physicalism and antireductionism generate irresolvable puzzles. Physicalists must therefore abandon antireductionism about biology and, for that matter, all of the special sciences. The argument begins with the physicalism that all parties claim to embrace. If the physical facts fix all of the facts, including higher-level ones, then the processes operating at the level of the purely physical are causally and constitutionally sufficient for all physical occurrences, including the biological ones, of course. So if biology reports that independent, irreducible causal processes operate at higher levels of organization to bring about biological events, states, and conditions, then these latter must be "overdetermined"; that is, they must have two or more independent causes, each of which would suffice in the absence of the other to bring them about and both of which actually always obtain: their full physical causes and some nonphysical ones as well. The only way to avoid this radically implausible and scientifically gratuitous conclusion is to hold that the physical and the biological causes are identical. But if this is the case, the argument goes, biological properties – the ones cited in (approximately) correct biological explanations – must be physical properties (or complex combinations of them, as we shall see). Thus, the

---

[1] A terminological point is worth making here and repeating occasionally in what follows. Reductionists do not advocate *reductive* explanations and theories. "Reductive" is a term of abuse employed in many disciplines to stigmatize explanations for being too simplistic, ignoring causally important factors and variables, unwarrantably reducing their number, or conferring excessive weight to one or a small number among them. For example, explanations of intelligence that appeal to genetic factors alone are often stigmatized, rightly, as "reductive." A reductionistic explanation of intelligence will appeal to all causally relevant factors, genetic and nongenetic, but it will insist that all of these factors operate through macromolecular processes.

biological domain is (metaphysically) reducible to the physical domain. Ergo, physicalist antireductionism is not a coherent view.

Accepting this argument's conclusion requires taking sides on a number of controversies in metaphysics not likely soon to secure widespread resolution. What is more, as noted, most reductionist philosophers, like their opponents, put more weight on how well their approach illuminates the actual character of biological practice than how it comports with fundamental philosophical commitments.

Biologists, of course, are interested in still a different matter – methodology. They take sides in the debate, whether they intend to or not, depending on the methodological choices they make. They are more likely to describe themselves as "opportunists," employing whatever method seems to work in a given inquiry. Nevertheless, those who seek, await, expect, or begin to craft molecular tools and explanations have accepted an inductive argument for reduction from previous successes and/or a philosophical one for it from metaphysics and epistemology.

The philosophical arguments for and against reductionism have persisted at least since the end of logical positivism in the early 1960s. They arose and became interesting as the positivist prohibition on metaphysical issues weakened. Among the matters on which philosophy had been silent for a generation that again became legitimate subjects were the nature, direction, and levels of causation, the relationship between the mind and the brain, and the realism/instrumentalism dispute regarding scientific theories. To all of these reductionism and antireductionism were related, as we shall see. The dispute about the aims of scientific theories was of immediate importance in motivating interest in reductionism.

Realism is roughly the thesis that the aim of science is to describe and explain the observable and unobservable nature of reality and that the sciences' explanations of observations by appeal to the unobservable or theoretical are to be taken literally and are apt to be tested and confirmed by the observations. Instrumentalism holds that scientific theories are merely instruments of organizing our expectations about predictions and that they are neither true nor false, or, making a slight concession, their truth or falsity is not material to their adequacy in the only task of science: prediction. Which theory and which tool one should employ to make predictions will depend on one's interests, needs, and resources. For example, one may well surrender improvements in precision provided by a molecular model for the greater economy of a "higher-level" model without ever taking sides on which one is (more) correct or whether they conflict in any way. Like saws and hammers, theories have different uses. One need not choose between them or order them in a hierarchy of being more or less fundamental.

It should be clear that the dispute about reductionism in philosophy presupposes substantial agreement that realism is the right view of theories in general. Both reductionists and antireductionists need to take the truth-values of theories seriously and agree that some theories actually do explain – including ones that advert to the theoretical – unobservable entities. Their dispute is supposed to be about the nature of and relations between these explanations, the kinds of regularities they appeal to, and the way in which the theories' "natural kinds" – their proprietary taxonomies – link up to the way nature is actually divided.[2] Antirealists have no reason to take sides on the reductionism/antireductionism dispute, as they hold that many competing and indeed contradictory theories may be equally adequate for one or more of the multifarious aims of science (which do not include revealing the truth about reality – i.e., realism). Arguments for antirealism about science, including instrumentalism, may appeal to the denial of reduction as a premise, but arguing against reductionism from antirealism as a premise is likely to be question-begging.

The term "natural kind" is a significant one in these debates. It has its origins in the notion that some concepts, categories, terms, and labels successfully "divide nature at the joints," in Aristotle's felicitous phrase. Science advances as it constructs the terminology that reflects these kinds. Mendeleev's periodic table of the elements is a clear example of the sort of advance achieved by finding the real natural kinds of a domain. A generation after Mendeleev, science came to know that the periodic table is the right taxonomy of chemistry when atomic theory provided a reductive explanation of its structure. Finding the natural kinds in any domain is a matter of empirical science, but philosophers have sought at least some necessary conditions for a term to qualify as a natural kind. The most obvious and unhelpful is the requirement that to be a natural kind a property must be a purely qualitative one that figures in real laws of nature – strict universal regularities of the sort with which chemistry and physics are replete. But that means a science without its own proprietary laws will also lack natural kinds. As we will see, this requirement is difficult for biology to satisfy, owing to its employment of the Linnaean taxonomy of genus

---

[2] The notion of a natural kind is closely connected to the concept of "levels" – higher and lower, as this notion is employed in debates about reductionism. No one has explicated the concept of "levels in nature" to general agreement. For the purposes of the present discussion, it will suffice to identify as a level of organization of nature a domain characterized by its own natural kinds expressed in its own proprietary laws. The same notion of levels will be important in the subsequent discussion of mechanism, although some mechanists have qualified or substituted an epistemically relativized notion of "levels" (Craver, 2015).

and species and to the role of Darwinian processes. The impact of this difficulty on arguments for and against reductionism will be considerable.

Although many scientists continue to employ the label of "reductionism" for a research program and methodology that they consider to have been vindicated by a range of explanatory achievements, especially in molecular biology, few philosophers have adopted the label, endorsed the scientists' claims, or sought to systematize them. However, by the end of the twentieth century, developments in psychology, especially in cognitive neuroscience, began to make the substance of reductionism increasingly attractive to philosophers. Under a new name of "mechanism," a new generation of increasingly prominent philosophers began to expound and advocate a systematic methodology to drive a research program in the life sciences. It is one that their philosophical predecessors would have recognized as reductionism "with a human face." The work of these latter-day reductionists, operating under the name of "mechanists," is the explicit subject of the last section of this Element. This Element will argue that mechanism is the continuation of reductionism's program by other means.

1 What Was Reductionism?

Debates about reduction can be dated from Earnest Nagel's treatment in *Structure of Science* (Nagel, 1961) and Hillary Putnam and Paul Oppenheimer's defense of reductionism in "The Unity of Science as a Working Hypothesis" (Oppenheim & Putnam, 1958). The combination of a commitment to a shared methodology among the sciences and a systematic hierarchy among them had been in eclipse since the time of Mill and Comte in the mid-nineteenth century. These post-positivist philosophers held that achievements first in physics, especially thermodynamics, and chemistry, particularly stoichiometric equation balancing, could not be understood without an explicit account of reduction. Owing to their distaste for metaphysics, these logical positivist and post-positivist philosophers restricted their reductionism to an explanatory and epistemological thesis only. Once philosophy of science shook off the shackles of logical positivism, philosophers became free to ground methodology and epistemology in claims about the fundamental nature of reality (i.e., metaphysics), and some did so.

1.1 Reductionism in Twentieth-Century Philosophy of Science

By the time philosophy of biology got started in the early 1970s, after the achievements in molecular genetics of Watson and Crick especially, those advocating reduction in biology could take inspiration from a reading of the history of the physical sciences from the seventeenth century onward. This was a chronicle of one apparent reductionistic success after another, beginning with Newton's derivation of Kepler's laws, through the Laplacian mechanization of the world, to Kelvin's assimilation of heat to mechanical motion, and ultimately Einstein's absorption of Newton's mystery of mysteries – gravity – to the curvature of space. Along the way, special relativity also made Newtonian mechanics a special case, deducible from more basic theory by setting c (the speed of light) equal to infinity and rest mass equal to mass in motion.

Reduction was initially introduced as an inter-theoretical relation between theories. Broader or deeper theories explained narrower theories as special cases or approximations. Post-positivists held that explanation in general was deemed to be a matter of deductive derivation from laws (whence the label "deductive-nomological [D-N] model of explanation"). The thesis was motivated by the recognition that most scientific explanations were causal, a metaphysically inspired insight that had to remain unexpressed during the hegemony of positivism and that remained unspoken in part because a few explanations, such as those provided by the ideal gas laws, did not show the temporal succession characteristic of causes. Even after the abandonment of the D-N model, the recognition of the importance of laws to sustain causal explanation implicitly or explicitly remained. The philosophy of science continued to honor David Hume's insight that causal claims are implicitly general and recognized the role of laws (often operating "behind the scenes") in the ability of causal claims to support counterfactual conditionals, as well as that explanatory power was concomitant with the support of such conditionals. Supporting a counterfactual conditional turned out to be a litmus test for being a law strong enough to explain. But which counterfactuals are true is controversial, and what the truth of a counterfactual consisted in was shrouded in mystery. This proves a vexatious complication for the reductionism debate, as we shall see.

Even as the D-N model was abandoned for one that deemed explanation a matter of identifying causal difference makers, inter-theoretical reduction continued to be viewed as derivation from laws, for that is what theories were taken to be: sets, indeed axiomatized sets of laws. And reductionism's *locus classicus*, Nagel's *Structure of Science* (1961), continued to be invoked. Nagel set two requirements for successful reduction: (1) The laws of the reduced theory must deductively follow from the laws of the reducing theory; and (2) the terms and concepts of the reduced theory must share meanings with the terms of the reducing theory. Although often stated explicitly, this second requirement is actually redundant, since valid deductive derivation presupposes the univocality of the language in which the theories are expressed. However, as exponents of reduction noted, the most difficult and creative part of a reduction is establishing these connections of meaning (i.e., formulating "bridge principles," "bilateral reduction sentences," or "coordinating definitions"). The ideal gas law, $PV = nRT$, was reductively explained by (derived from) the kinetic theory of gases only after Kelvin realized that heat can be defined in terms of the mean kinetic energy of the constituent particles comprising gases. Establishing the bridge principles required by a reduction is among the greatest of theoretical achievements, even if once stated they are tantamount to definitions. Although a "bridge principle" may only take the form of an equation (e.g., $T = \frac{1}{2}mv^2$), its physical significance is much greater. It economically expresses the profound metaphysical reduction of more than one kind of thing in one whole realm of phenomena – that of thermodynamics – to another more fundamental, more basic kind of thing – mass in motion, a domain already well systematized by Newtonian mechanics. Similarly, in biology, Watson and Crick's achievement consisted in telling the world what genes really are by showing how polynucleotide sequences can do the work of replication and phenotypic regulation required of genes as units of heredity. Expressing the achievement of identity explicitly in these bridge principles is what the second requirement of reduction demanded, so it was worth stating independently of the requirement of deduction that makes it logically redundant.

The deduction/derivation approach to the reduction of theories needs some qualification and articulation in light of the advent of "modeling" in science and its treatment by philosophers of science. By the late 1960s, it became evident to philosophers that the unit of scientific theorizing had increasingly departed from the syntactically characterized deductive systems that had been used for theories in the period of logical positivism. Instead, theoretical advance had in many disciplines, and especially biology, become a matter of the development of models – representations of the dependence of observable variables on other, often unobservable ones. But modelers often knowingly simplified some features, exaggerated others, and neglected some altogether. Models were almost never modeled by philosophers of science as sentences that could be arranged in axiomatic systems whose components included underived and derived laws that allowed for deduction. What is more, models clearly had other functions than explanation and, when employed for that purpose, could not honor the D-N model.

The advent and importance of modeling as the preferred vehicle of theoretical development did not pose insuperable obstacles to the "spirit" or even the most important details of the post-positivist treatment of reduction. Putting aside all of the other uses of models, consider only their role in explanation and the testing that corroborates explanation. Following Nagel, reductionists would hold that science progresses by the development of models that successively explain an older, more limited model's successes by showing that the older, less detailed, or less general model's character could be recovered from the newer, more detailed, or more general model by restricting the ranges of some of its parameters and variables. This captures the derivational relation that reduction demands. Moreover, and for many of the same reasons, the succession of models in a discipline with overlapping explanatory and predictive domains also requires that concepts, properties, and kind terms of simpler models be linked systematically to the distinctive variables of the newer models. Otherwise, their domains would not overlap. Focusing on models as representations of reality clearly presupposes a commitment to realism about science's goal: constructing an accurate representation of reality. Reductionism versus antireductionism is a dispute that is immediately and unavoidably raised by this commitment: Are there metaphysically autonomous layers or levels of reality or not? The dispute about the character of theories as models or axiomatic systems turns out to be largely orthogonal to metaphysical and explanatory issues in the reductionism/antireductionism dispute.

Almost immediately after its articulation in the late 1950s, problems arose for the beautifully sweeping account of scientific progress as reduction. First, philosophers noted difficulties accommodating biology to its demands. Then, it turned out that reduction in physics was a much more complicated matter than initially supposed.

It was the philosopher of biology Kenneth Shaffner who first noted in 1967 that reduced theories are usually less accurate and less complete in various ways than reducing theories, and therefore are incompatible with them in predictions and explanations – Nagel (1961) had also mentioned it in passing. According to Shaffner, a further, third requirement on reduction needed to be added: The reduced theory had to be "corrected" before its derivation from the reducing theory could be effected. In biology, this is obvious: Mendel's "laws" need to be corrected to accommodate crossover and linkage before they can be derived from the molecular details of meiosis. But in physics, too, this is required: Newton's theory and relativistic mechanics disagree – they are logically incompatible, so the former can hardly be derived from the latter. This raised a problem that became nontrivial in the fallout from Thomas Kuhn's *Structure of Scientific Revolutions* (1961) and Paul Feyerabend's *Reduction, Empiricism and Laws* (1964).

It became evident in these works that "correction" sometimes resulted in an entirely new theory. Even after "correction," when derivation from the reducing theory is possible, this showed nothing about the relation between the original pair of theories. Feyerabend's examples were Aristotelian mechanics, Newtonian mechanics, and relativistic mechanics, whose respective crucial terms "impetus" and "inertia," "absolute mass," and "relativistic mass" could not be connected in the way reduction required. The "natural kind" terms of the reduced and reducing theories were distinct and not inter-definable, even when the same word or symbol figured in both theories, such as "mass." The problem of linking natural kinds in biology recurs in the difficulty of connecting the geneticist's term "gene" with the molecular biologist's term "gene," as David Hull was the first to point out in 1974. This is the problem of multiple realization that Hull recognized in the redundancy of the genetic code and that was amply magnified by much subsequent research on how DNA realizes the gene: The same functional gene can be equally well realized by any of a large disjunction of different primary polynucleotide sequences, sequences in which the replacement of one of the four nucleotides that comprise DNA molecules makes no difference to their biological activity. As there are often hundreds or thousands of nucleotides in a gene, listing all of the combinations that are biologically equivalent is impossible and biologically insignificant. As we will see, the difficulty Hull was the first to note unravels the entire project of reduction as the straightforward deductive derivation of narrower theories – even corrected ones – from broader ones, especially in the life sciences.

The suggestion that the history of scientific "progress" is a matter not of reduction but of *replacement* emerged with Kuhn's hugely influential work. *The Structure of Scientific Revolutions* argued that theories, the research programs they underwrite, and indeed the conceptions of nature they articulate are not assimilated into their scientifically superior successors. Instead, they are replaced by new *paradigms* in conceptual revolutions that reframe their disciplines without preserving past achievements. What's more, as Kuhn claimed to document, the institutional discipline that science imposes requires that its history be regularly rewritten after each replacement – each paradigm shift – in order to make these replacements easy to mistake for mere corrections and smooth reductions. This mythic self-deception was required, according to Kuhn, by the institutional imposition of disciplinary norms in each of the sciences. In *The Structure of Scientific Revolutions*, Kuhn explicitly rejected the reductionist's vision of science as providing an increasingly accurate description of the nature of reality. Instead, it was one socially "constructed" paradigm after another.

By contrast with Kuhn, Feyerabend, and their followers, the first reductionists and their latter-day adherents endorsed the strongly progressive vision of the history of science as cumulative in at least two different ways, both of which vindicated reductionism: diachronic and synchronic reduction. (The distinction was first drawn by Nickles, 1973.) Diachronic reduction obtains when earlier theories are shown to be simplifications of or approximations to their successors. These newer theories vindicate the older ones, at the same time explaining their approximate explanatory and predictive success while also explaining exactly where and why they failed in prediction or explanation. Later theories show earlier ones to be *special cases* that neglect some variables, fail to measure certain coefficients, or set parameters at restricted values.

Diachronic reduction is exemplified by the relation between Newtonian mechanics and relativistic mechanics. Philosophers advocating reduction as an account of the secular progress of science had to deal with the difficulty of defining Newtonian terms in relativistic ones, which had motivated the Kuhnian denial that science cumulates.

The other sort of reduction – synchronic – is far more important to biology. It goes back to Comte's nineteenth-century picture of a hierarchy of sciences from more to less fundamental. It was also Shaffner who coined the term "layer-cake reduction" to reflect the latter-day version of this notion that less fundamental theories are to be explained by reduction to more fundamental theories: At the basement level rests some unification of quantum mechanics and the general theory of relativity, above these physical and organic chemistry, then molecular biology and functional biology, and at the higher levels are neuroscience, psychology, economics, and sociology. Synchronic reduction is supposed to be explanatory because on the account of explanation associated with reduction – the D-N model – explanation was logical deduction, and the explanation of laws required the deduction of laws from other laws. Synchronic reduction obviously also involves mereological decomposition, in which the behavior of more composite items described in reduced theories is explained by derivation from the behavior of their components as described by the reducing theory. As we shall see, synchronic reductionism was eventually transformed into a thesis that labeled itself "mechanism" and became a widespread doctrine in twenty-first-century philosophy of science, and especially among philosophers of biology and neuroscience.

Most philosophers of science share a commitment to science as epistemically progressive, as securing more and more knowledge about a wider and deeper domain, which these two forms of reductionism – diachronic and synchronic – endorsed. But they increasingly dissented from the explanations offered for this fact about science by the account of reduction as the deductive derivation of theories.

1.2 The Philosophy of Biology Takes Off

Soon after the philosophy of biology "took off" as a distinctive domain of research in the philosophy of science, two facts about the domain of biology became evident that completely undermined reduction as derivation: First, as philosophy of science *then* understood the term "law," there are no laws in functional biology to derive from molecular biology, nor laws in molecular biology to derive from biochemistry, organic chemistry, or any other kind of chemistry; and second, the proprietary (presumptive natural) kinds of biology – its vocabulary of functional properties – couldn't be defined in the terms of molecular biology, and so it couldn't be reduced to chemistry. And these two obstacles to reduction by derivation had the same biological sources. Consequently, either reductionism about biology had to be surrendered or it had to be radically reconfigured.

Recall the requirement for deductive derivation that the terms, concepts, and distinctive vocabulary of the reduced theory must have the same meaning in the reducing theory. This requirement is redundant as it draws attention to a condition already made necessary by the first requirement of deduction. Without it, derivations are unsound. Consider the deductively valid argument God is love, love is blind, therefore God is blind. This is unsound owing to the equivocation on "blind." Sameness of meaning between theories is to be assured by "bridge principles," often themselves important achievements such as Kelvin's equation of temperature with mean kinetic energy. But, as Hull first noted in 1974, the required "bridge principles" between the concept of gene as it figures in functional biology (population biology, evolutionary biology, developmental biology) and as it figures in molecular biology could not be constructed. Any attempt to define, say, the gene for hemoglobin in terms of a nucleic acid sequence is bound to be incomplete. To begin with, as Hull noted, the genetic code is redundant, so that three different DNA nucleotides can constitute the same codon for a given amino acid and different codons can code for the same RNA sequence and same amino acid. So, the set of nucleotide sequences that are each of them an instance of the gene for hemoglobin will be vast, and certainly not worth actually enumerating for any scientific reason. Moreover, if the genetic code is a frozen accident, in Crick's words, there will be sequences never actually realized that could code for the same RNA sequences and the same functional protein. A theoretically well-grounded bridge principle would need to include these hitherto unrealized alternatives. Otherwise, it would not support counterfactual claims about biologically possible (but nonfactual) sequences constituting a particular gene. As such, it would not be a bridge principle of the required sort, but merely a laundry list of alternative sequences that had been realized in the past but might not have been realized then and might be superseded by other sequences in the future. The possibility is biologically far-fetched perhaps, but it is a problem for reductionists who held that bridge principles are important theoretical, explanatory achievements.

But matters were also more complicated biologically than anyone realized in 1974. Subsequent discoveries in molecular genetics made the provision of usable or even finitely statable bridge principles linking genes and DNA quite impossible. Once introns and exons were discovered, the question arose as to whether the latter were parts of the genes from which they were removed during or after transcription. This complicated the construction of bridge principles further. Then it was discovered that the same nucleotide sequences can code for different gene products and therefore constitute different functional genes. The molecular biologists' identification of promoters and suppressors – *cis*- and *trans*-acting regulatory sequences operating on structural sequences – further increased the obstacles to finding any manageable systematic link between what the molecular geneticists' were discovering about nucleotide mechanisms and the functional biologists' explanatory concept of the gene. Satisfying the requirement of any conceptual link between molecular biology and functional biology that would vindicate deductive explanatory reduction was clearly never going to happen. Their natural kinds just didn't line up.

The upshot of Hull's problem – and its vindication by developments in the science after his original insight – is grave. Two problems arise for metaphysical reductionism (i.e., physicalism), the thesis everyone thought was safely beyond dispute. First, if physical facts fix all of the facts, then facts about genes need to be fixed by facts about nucleic acids. But without an account of how this is actually done, reductionism is vulnerable to the charge of "hand-waving" dogmatism. At a minimum, metaphysical reduction, also known as physicalism, needs a good non-question-begging explanation of why no complete account is available of how physical facts fix biological ones. Second, if possible, this account should be combined with a nondeductive conception of reduction, one that still shows how molecular biology can and does explain functional biological phenomena. And finally, these explanations are going to have to be *improvements* on nonmolecular explanations in functional biology if the research program of reductionism is to be vindicated. Leading molecular biologists supposed that their research was reducing "the gene" to nucleic acid sequences and gene expression to molecular biology. It was up to the philosophers who agreed with them to show how they were doing this. Deductive derivation was not the way. (Notice also that this problem bedevils biological models as much as axiomatic theories.)

The second problem facing reductionism in biology was the almost complete absence of strict laws at the level of either the reducing theory or the reduced theory. If there aren't any laws in either theory, there is no scope for derivational reduction at all.

The notion of a "strict law" will be important hereafter, so it would be well to "explicate" it. A strict law is a regularity of the form "$(x)(Fx \rightarrow Gx)$," "All A's are B's," or "Whenever C occurs, E occurs." Examples of such strict laws are arguably to be found in physics, such as Newton's laws, including the inverse-square law of gravitation, or in chemistry, such as various stoichiometric regularities as in "$2H_2 + O_2 \rightarrow 2H_2O$." Regularities that express these universal relations are "strict" by contrast to *ceteris paribus* regularities that hold only "other things being equal," where this clause excludes a usually indefinite number of conditions, some or many of which are unknown. Some philosophers have argued that all laws contain explicit or implicit *ceteris paribus* clauses, including even the most fundamental laws of physics (Cartwright, 1983). On this view, there are no "strict laws" in science. Newton's gravitational force law may be said to hold only *ceteris paribus* in so far as other physical forces are not also in operation. But they always are in operation, so this law doesn't hold only *ceteris paribus*, it doesn't ever actually obtain at all. In what follows, the expression "strict law" will be employed to accommodate this claim by noting that the proprietary laws of a domain operate, obtain, and regulate phenomena in the domain only together with the other proprietary laws governing the domain. Their "writ" will operate only *ceteris paribus*, but the conditions under which they do so will be limited, known, and identified by the other proprietary laws of the domain. The laws of physics and chemistry are strict in this sense. Or rather, our best hypotheses about what these laws are suggest that they are strict in the way here qualified.

Understanding the reason why there are no strict laws anywhere within the various subdisciplines of biology (i.e., beyond the theory of natural selection) is not only essential for understanding why post-positivist reductions are impossible; addressing this problem and diagnosing biology's lack of real laws turns out also to solve the physicalist's first problem of reconciling physicalism with the irreducibility of natural kinds. It also explains why there are no counterfactual supporting bridge principles connecting the descriptions of functional biology to the concepts of molecular biology.

That there are no strict laws in biology, only relatively restricted *ceteris paribus* regularities, is now widely recognized among philosophers of biology. The reason why there are no strict laws in biology is largely biological, not philosophical, and this is one reason why the conclusion has achieved some consensus among philosophers. But the *ceteris paribus* regularities of biological science play much the same explanatory role in biology that strict laws play in physical science. So, many philosophers of biology began to insist that these *ceteris paribus* regularities are laws as well, or at least fill the explanatory function of laws in biology. In effect, these philosophers (Sober, 1993; Lange,

1995; Mitchell, 2000) revised and widened the concept of law. In doing so, they opened a whole new agenda of questions about how biological explanation works, since *ceteris paribus* regularities had long been suspect in science and philosophy. Suspicion of *ceteris paribus* regularities had existed at least since the logical positivists, initially due to their untestability and later due to doubts about whether they support counterfactual conditionals in the way required for a generalization to have explanatory power.

The prominence of *ceteris paribus* regularities in biology stems from the fact that many of its explanatory concepts – its proprietary kinds – and almost all of its descriptive concepts appeal to functions. For instance, to call something a wing, a fin, or a gene is to identify it in terms of its function. It was Darwin's achievement to show how functions arise and persist in the biological domain by showing that they are all of them adaptations. Biological functions are naturally selected effects (Wright, 1976).[3] However, it is crucial to recognize that natural selection for adaptations (i.e., environmentally appropriate effects) is blind to differences in physical structure that have the same or roughly similar effects. Consider the approximately 40 different ways in which Darwinian processes have realized members of the class of organs that are wings, or the 4 different ways it has constructed the eye. Whatever physical or structural property all wings or eyes have in common will be due to homology or constraint dictated by physical law. Even among vertebrate wings, heterogeneity of structure rules: consider the differences between bat wings, bird wings, and flying fish wings, not to mention pterodactyl wings.

Natural selection "chooses" variants by *some of their effects* – those that fortuitously enhance survival and reproduction. When natural selection encourages variants to become packaged together into larger units, the adaptations become functions. Selection for adaptation and function kicks in at a relatively low level in the organization of matter. As we'll see in Section 3, when molecules develop the disposition chemically, thermodynamically, or catalytically to encourage the production of more tokens of their own kind, natural selection comes into force. To employ the vocabulary of Dawkins (1982) and Hull (1989), at this point in the aggregation of matter, replicators and interactors (or vehicles) first appear. As a result of purely physical

---

[3] Dissent from this view arose among philosophers of biology who embraced Cummins's (1975) notion of function. On Cummins's account, a function is a causal capacity that is nested in a broader system. This broader system engages in the "programmed manifestation" of some more complex capacity. The original component's function is the causal contribution it makes to the manifestation of this more complex capacity, relative to some "analytical account." Advocates of the selected effects account of functions argue that all such causal capacities in biology are the result of natural selection, which provides the litmus test for whether a causal capacity is a biological function (see Neander & Rosenberg, 2009).

processes, some molecules become replicators. They template, catalyze, or otherwise encourage the production of copies of themselves. These copies then interact with the environment so that changes in them – mutations – will result in changes in their rates of replication in their environments. Among such replicating and interacting molecules, there are frequently to be found multiple *physically distinct* structures with some (nearly) identical rates of replication, different combinations of different types of atoms and molecules that are about equally likely to foster the appearance of more tokens of the types they instantiate. This structural diversity explains why no simple identification of molecular genes with the genes of population genetics (of the sort derivational reduction requires) is possible. More generally, the reason why there are no laws in biology is thus the same reason why there are no bridge principles of the sort post-positivist reduction requires.

It is the nature of any mechanism that selects for effects that it cannot discriminate between differing structures with identical effects. And functional equivalence combined with structural difference will always increase as physical combinations become larger and more physically differentiated from one another. Moreover, perfect functional equivalence isn't necessary. Mere functional similarity will do. Since selection for function is blind to differences in structure, there will be no laws in any science that, like biology, individuates kinds by selected effects – that is, by functions. A law in functional biology will have to link a functional kind either with another functional kind, such as "All butterfly wings have eyespots," or a structural kind, such as "All eyespots are composed of proteins." But neither of these statements can be a strict law because of the blindness of natural selection to structure.

It is crucial to see exactly why this is so – why biology's functional kinds are always too heterogeneous in structure and composition to allow either for bridge principles linking function to structure or for strict laws at the level of function at all.

To see why, consider the form of a generalization about all F's, where F is a functional term, like gene, cell, wing, or heart. The generalization will take the form $(x)(Fx \rightarrow Gx)$, a law about F's and G's. If Gx is a structural predicate, the generalization can serve as a bridge principle. If Gx is a functional predicate, the generalization will be a purely functional regularity.

Consider the first possibility. If G is structural, then G will describe some physical attribute common to its instances. For example, all genes (a functional kind) are composed in accordance with the 64-unit nucleic acid codon code (their common structural character). To begin with, this regularity reflects what most molecular biologists, following Crick, called a "frozen accident." Its character as a local and only temporarily true regularity was revealed in a particularly striking way when a variety of synthetic codon codes were engineered over a period after the year 2000. As a result, many counterfactuals about the genetic code turned out to be unsustainable.

The very forces that made the genetic code a local regularity – natural selection – can unmake it (and have done so already if we count human interventions as evolutionary forces). If this process operates at the fundamental level of molecular biology, then its operation elsewhere in the biological domain can never be ruled out. Thus, natural selection makes any proposed bridge principle that grounds a functional kind in a structural kind at most a local regularity, reflecting just more temporarily frozen accident, like the genetic code. Anatomical and physiological reduction, for example, won't just be a matter of finding bridge principles that connect higher-level, derived laws to lower-level, more basic laws at all.

Consider the second alternative. Suppose that the regularity we seek to explain or to employ as a bridge principle relates two functional types, categories, or properties, instead of a functional kind to a structural one. We seek a regularity about a second, different function carried out by all instances of a certain functionally described class of objects. It should be obvious that there will be no such regularity that combines truth with nontriviality and scientific significance. At best, if true, such regularities will have the trivial character of the statement that animals with wings fly. Leaving aside the many winged creatures that do not fly at all, there are a number of different ways in which flight can be accomplished by wings of different types: hovering as hummingbirds and flying insects or exploiting aerodynamic principles in the different ways bats and birds do, not to mention the approaches of flying fish and pterodactyls. It's clear that there are too many different types of wings to ground the regularity's antecedent kind "wing" or "animal with a wing" in a finitely manageable disjunction of different ways to realize wingedness, and equally too many ways physically to realize the consequent's kind "flight," for the generalization, even suitably qualified, to be explained by derivation from some physical theory of aerodynamics.

In general, the existence of another functional property different from F that all items in the extension of the functional predicate Fx share must be highly improbable. If Fx is a functional kind, then owing to the blindness of selection to structure, the members of the extension of Fx are physically diverse. As such, any two F's will have nonidentical (and usually quite different) sets of effects. Without a further effect common to all F's, selection for effects cannot produce another selected effect; it cannot uniformly select all members of F for some further adaptation. Thus, there will be no further function kind that all F's share in common.

Biological, functionally characterized traits, features, organisms, and components will of course all be physical. So, being physically constituted is one structural feature that will be shared by all of the members of any functional kind F, and there may be other, more specific physical ones as well. But these will not be biologically interesting features. Rather, they will be properties shared with many other things, such as mass or electrical resistance, properties that have little or no explanatory role with respect to the behavior of members of the extension of Fx. For example, that the generalization that "all mammals are composed of confined quarks" does relate a structural property –quark confinement – to a functional one – mammality – is exceptionlessly true, but it is not a law of biological interest. Any science in which kinds are individuated by naturally selected effects will have few if any exceptionless laws.

Ever since Darwin's focus on artificial selection, it has been recognized that in the evolution of some species, other species constitute the selective force channeling their genetic changes. The interaction of predator and prey manifests the same relationship. Since the importance of frequency-dependent selection became apparent, it has been recognized that an interbreeding population can be an environmental force influencing its own evolutionary course. Competition for limited resources is endemic to the biosphere. Any variation in a gene, individual, line of descent, or species that enhances fitness in such a relentlessly competitive environment will be selected for. Any response to such a variation within the genetic repertoire of the competitor gene, individual, lineage, or species will in turn be selected for by the spread of the first variation, and so on. One system's new solution to a design problem is another system's new design problem. If the "space" of adaptational "moves" and countermoves is very large and the time available for trying out these stratagems is long enough, every regularity in biology about functional kinds will be falsified (or turned into a stipulation) eventually.

What this means, of course, is that any functional generalization in biology will be a *ceteris paribus* generalization in which, over evolutionary timescales, the number of exceptions will mount until its subject becomes extinct. Take a simple example, such as "butterflies have eyespots." The explanation for why they do is that eyespots distract birds from butterflies' more vulnerable and more nutritious parts and provide camouflage by giving the appearance of the eyes of owls that prey on birds. This stratagem for survival can be expected in the long run to put a premium on the development of ocular adaptations among birds – say, the power to discriminate owl eyes from eyespots – that foil this stratagem for butterflies. This in turn will lead either to the extinction of butterflies with eyespots or the development of still another adaptation to reduce predation by birds – say, the development of an unappetizing taste or a shift in color to the markings of another species of butterfly that already tastes bad to birds. And in turn this stratagem will lead to a counterstroke by the bird lineage. The fantastic variety of adaptational stratagems uncovered by biologists suggests that there is a vast space of available adaptive strategies among competing species, and that large regions of it are already occupied. The upshot is that to the extent that general laws must be timeless truths to which empirical generalizations approximate as we fill in their *ceteris paribus* clauses, no such strict laws are attainable in biology because we can never fill in these clauses.

Notice that this result obtains as much for molecular biology as it does for functional biology. Because the kinds of molecular biology are also functional, even at the biochemical level, natural selection's persistent exploration of adaptational space makes for lawlessness at the level of macromolecules as well. Consider three examples of generalizations in molecular biology that were once held to be strict laws and now have been found to have exceptions: All enzymes are proteins; hereditary information is carried only by nucleic acids; and the central dogma of molecular genetics – DNA is transcribed to RNA and RNA is translated to protein. It turns out that RNA catalyzes its own self-splicing, that prions (proteins responsible for mad cow disease) carry hereditary information, and that retroviruses carry their own hereditary material in RNA and transcribe it to DNA. These exceptions to the relevant generalizations emerged through the operation of natural selection – finding strategies in adaptational space that advantage one or another unit of selection in the face of stratagems employed by others.

It is worth reiterating that, as many philosophers of biology will argue, biological explanation does not require strict laws. Non-strict *ceteris paribus* laws will suffice for biological explanation. Models, for example, apply to systems only *ceteris paribus*. But the issue here is not (yet) explanation – it is the metaphysical thesis of reduction and the methodological dictum to seek reductions. Without strict laws, the derivational demands of metaphysical and methodological reductionism are impossible to act upon. Without recourse to laws of either kind, reductionism must be rejected or reformulated. Any reformulation of reductionism needs to be reconciled with the explanatory power of those *ceteris paribus* regularities (and/or the claims that models explain by applying *ceteris paribus*). And the explanatory power of these *ceteris paribus* regularities and models will have to accommodate the fact that, owing to Darwinian natural selection, neither the regularities nor the models can support counterfactuals beyond very local circumstances, if at all.

The absence of laws in biology makes untenable the derivational account of reduction, and in doing so also undermines the post-positivist account of the unity of scientific results and the unity of scientific methods of securing these results. Unlike physics and chemistry, the research program of the life sciences will not consist in the discovery of the sort of universal nomological regularities that reveal the real natural kinds and the laws of working that explain the kind's instances. Metaphysical reductionism will have to seek a new rationale in the absence of the derivational account of nomological explanation and the research program of seeking laws of nature in biology.

1.3 Twenty-First-Century Antireductionism

Antireductionism labels the claims of philosophers and others who rejected any sort of reduction in biology as metaphysics, as explanation, and as methodology. These philosophers and several important biologists who made common cause with them endorsed several positive claims about biology: (1) There are generalizations at the level of functional biology; (2) these generalizations are explanatory; (3) there are no further generalizations outside of functional biology that explain the generalizations of functional biology; and (4) there are no further generalizations outside of functional biology that explain better, more completely, or more fully what the generalizations of functional biology explain.

It's worth repeating here that these four claims can be formulated in terms that reflect the search for models in biology. Antireductionism is equally well expressed as the thesis that there are models in functional biology that explain a range of cases adequately, that their explanatory power cannot be enhanced by the provision of models in molecular biology, and indeed that the phenomena that the higher-level models explain cannot be explained by models operating at lower levels.

Whether expressed as a claim about generalizations or models, all four components of antireductionism are challenged by at least some of the same problems that vex reductionism, for, as is obvious, each of them requires that functional biology comports explanatory generalizations of the very sort whose absence makes derivational reduction impossible. But besides this problematic presupposition that antireductionism shares with reductionism, it has distinct problems of its own.

To see the distinct problems of antireductionism, consider a paradigm of putative irreducible functional explanation advanced in a justly famous antireductionist argument: P. Kitcher's "1953 and All That: A Tale of Two Sciences." Kitcher begins with a well-recognized regularity of functional biology, one of Mendel's laws updated:

> (G) Genes on different chromosomes, or sufficiently far apart on the same chromosome, assort independently.

The antireductionist proffers an *explanans* for (G), which we shall call (PS):

> (PS) Consider the following kind of process, a *PS*-process (for *pairing* and *separation*). There are some basic entities that come in pairs. For each pair, there is a correspondence relation between the parts of one member of the pair and the parts of the other member. At the first stage of the process, the entities are placed in an *arena*. While they are in the arena, they can exchange segments, so that the parts of one member of a pair are replaced by the corresponding parts of the other members, and conversely. After exactly one round of exchanges, one and only one member of each pair is drawn from the arena and placed in the *winners box*.
>
> In any PS-process, the chances that small segments that belong to members of different pairs or that are sufficiently far apart on members of the same pair will be found in the winners box are independent of one another. (G) holds because the distribution of chromosomes to games at meiosis is a PS-process.

He concludes:

> This I submit is a full explanation of (G), an explanation that prescinds entirely from the stuff that genes are made of. (Kitcher, 1984, pp. 341–343)

Although Kitcher doesn't call it a model, we might well treat (PS) as one that explains why (G) obtains, since it successfully models meiosis.

Leave aside for the moment the claim that (PS) is a full explanation of (G) and consider why, according to the antireductionist, no molecular explanation of (PS) is possible. The reason is basically the same story given above about why the kinds of functional biology cannot be identified with those of molecular biology. Because the same functional role can be realized by a diversity of structures and because natural selection encourages this diversity, the full macromolecular explanation for (PS) or for (G) will have to advert to a range of physical systems that realize independent assortment in many different molecular ways. These different ways will be an unmanageable disjunction of alternatives so great that we will not be able to recognize what they have in common, if indeed they do have something in common beyond the fact that each of them will generate (G). Even though we all agree that (G) obtains in virtue only of macromolecular facts, nevertheless we can see that, because of their number and heterogeneity, these facts will not explain (PS), still less supplant (PS)'s explanation of (G), or for that matter supplant (G)'s explanation of particular cases of genetic recombination. This is supposed to vindicate all four of antireductionism's theses: that functional explanations are adequate by themselves and that functional explanations cannot be explained further or more deeply by nonfunctional ones, nor can they be replaced by them.

But this argument leaves several hostages to fortune. Begin with (G). If the argument of Section 1.2 is correct, then (G) is not a law at all, but a *ceteris paribus* generalization; in fact, G will be the report of a long conjunction of a large number of particular facts about a spatiotemporally restricted kind, "chromosomes" of which there are only a finite number extant over a limited time period at one spatiotemporal region (Earth). Accordingly, (G) is not something that by itself is a target for derivational reduction to the laws of a more fundamental theory. If (G) as an updated version of Mendel's second law is accorded explanatory power (giving part of the explanation of the random distribution of Mendelian phenotypes), then the explanation of (G) is not going to look anything like the derivation that the D-N model demands, for it will have to include facts about the local terrestrial conditions on which natural selection worked to produce genes on chromosomes. Every explanation in biology is going to have to be "historical," making at least implicit appeal to the local environmental conditions that selected for the explanatory processes it adverts to. An explanation invoking (G), or more accurately (G *ceteris paribus*), doesn't vindicate the four theses that antireductionism advances about explanatorily autonomous generalizations in functional biology.

Biologists certainly do accord explanatory power to (G). But how does (G) explain if it is a merely local regularity, an accidental one put in place by natural selection operating on local circumstances? And the same questions arise about (PS)'s nonreductive explanation of (G). Thus, what certifies the account of PS-processes given above as explanatory? Antireductionists are also physicalists after all, and so they accept that there is a vast disjunction of macromolecular accounts (including chemical laws and boundary conditions described biochemically) of the underlying mechanisms of meiosis that make (PS) true. Why doesn't this disjunction of macromolecular accounts explain (PS) and thereby deepen (PS)'s explanation of (G) and indeed explain whatever it is that (G) explains?

Exactly why is it that, according to antireductionists, a full or complete macromolecular explanation of (PS) is not on the cards? That there is a disjunction of macromolecular pathways that implement PS-processes and thus bring about (PS) and (G) does not seem to be at issue. Only someone who denied the thesis of physicalism – that the physical facts fix all of the biological facts – could deny the causal relevance of some vast, motley, but finite number of disparate macromolecular processes to the existence of PS-processes and the truth of (G). Of course, no one could complete the list of all of the different actual and possible macromolecular pathways that PS-processes consist in, and it would be scientifically fatuous to do so in any case. But metaphysical reductionism doesn't require that humans can actually effect such a complete enumeration, nor does the epistemological thesis that reductive explanations deepen understanding. The demand that reductive explanation be deductive would require it, but that is a demand on explanations reductionism needn't – indeed, shouldn't – make.

Antireductionists such as Kitcher have ruled out one answer to the question of why all of the macromolecular pathways that constitute PS-processes don't explain them. This is an answer to the question that begins with the claim that scientific explanation must respect the cognitive and computational limitations of human beings. It is beyond the cognitive powers of any human even to contemplate the vast disjunction of differing macromolecular pathways that constitute PS-processes, let alone to see how each of them gives rise to meiosis. Accordingly, it is not in our powers to recognize that disjoined they constitute an explanation of (PS). Similarly, it is beyond the competence of biologists to recognize how each of these macromolecular processes gives rise to (G). Ergo, no one can provide anything even close to a complete reductive explanation of PS-processes or, for that matter, (G). This argument implicitly adopts theories of explanation that treat them not as reflecting causal relations in nature that obtain independent of us, but as being thoroughly conditioned by human interests and capacities. But, as noted, few antireductionists wish to base their claims against reduction merely on the claim that an explanation must be understandable by us in order to be adequate, for then antireductionism would be a claim not about biology, but about (possibly temporary) limitations on human cognitive powers; their antireductionist claim would be threatened with refutation by the emergence or the arrival of far more cognitively powerful agents than mere *Homo sapiens*. As Kitcher recognizes, antireductionism needs "a reply to the reductionist charge that we [the antireductionists] reject the explanatory power of the molecular derivation [of (G) and (PS)] simply because we anticipate out brains will prove too feeble to cope with its complexities" (Kitcher, 1984, p. 344). (For a different view regarding the role of human cognitive limits in scientific explanation, see Potochnik, 2017.)

In "The Multiple Realizability Argument against Reductionism," Elliot Sober (1999) observes another limitation facing arguments that the vast motley of molecular details obscures understanding and is therefore not explanatory. Such arguments assume that generality is always a virtue in explanations, even when it conflicts with demands for depth. Explanatory reductions, Sober notes, may sacrifice generality for detail and, given the interests of biologists, are in some circumstances to be preferred for doing so. This Solomonic judgment will satisfy neither party to the debate, since it is not about whether or not reduction's depth is sometimes to be preferred to the generality of higher-level models and generalizations. The question at issue is whether reduction always effects an explanatory improvement. (This matter will become more salient in Section 4 below.)

Reductionists will appeal to physicalism – the thesis that the physical facts fix all of the facts – to underwrite the thesis that every PS-process is a physical one. We may be confident that there is a truth to the form

(R) $PS = P_1, v\, P_2\, v \ldots v\, P, v \ldots v\, P_m$

where P_i is one macromolecular method by which pair separation (PS-processes) among chromosomes can be realized and m is the number – a very large number – of all of the multiple ways macromolecular processes can realize PS-processes. This biconditional will hardly work the same magic on our intuitive understanding of PS-processes that Kelvin's equation of temperature as mean kinetic energy achieved in showing that heat is molecular motion, but at least in principle it could do so. It will be a crucial step in elaborating the deep explanation for all PS-processes, and by doing so it will also ground such generality as explanations employing (PS) models may secure.

Antireductionists may accept that we can at least in principle formulate such an expression as (R). Their own allegiance to physicalism should require them to make this admission. But they will make two crucial observations about (R): Even if it describes all of the actual ways in which PS-processes are realized, it is incomplete in not identifying all of the *physically possible* ways in which they may be realized, and therefore it lacks the counterfactual strength required for explanatory power. Moreover, antireductionists will observe, there is after all something that the vast but finite disjunction of actual macromolecular realizations of (PS) have in common that would enable the disjunction of them fully and with generality to explain any particular instance of a PS-process to someone with a good enough memory for details. But rather than vindicating reduction, what they have in common is a feature that reveals the fundamental reason why the thesis fails for biology.

Each of the Pi's – the diverse physical realizations of (PS) – was selected for just because each implements a PS-process and therefore is adaptive in the local environment of Earth from about the time of the onset of the sexually reproducing species to their eventual extinction. Since selection for implementing PS-processes is blind to differences in macromolecular structures with the same or similar effects, there may turn out to be nothing else completely common and peculiar to all macromolecular implementations of meiosis besides their being selected for implementing PS-processes. But that might be quite enough for the antireductionist's claim, for it is open both to us and to creatures with cognitive powers vastly in excess of humans to grasp what it is about the disjunction of all of these macromolecular implementations that enables each to explain some PS-processes or others and thereby each case of (G): the process of natural selection that Darwin uncovered.

Because the Darwinian process carves functions out of physical processes by the mechanism of blind variation and natural selection, it unifies physically disparate processes and structures into all of the natural kinds of biology. These then figure in explanatory regularities and models. And since it operates even at the deepest level of detail in molecular biology, it makes irreducible the regularities and models of this subdiscipline, along with the natural kinds in which they are expressed and the explanations to which they are applied. The reason for this, antireductionists will insist, is obvious. The process of natural selection is itself one that operates at many levels in the biological domain, and *irreducibly* so.

2 Biology as Natural History

Fifty years of work in the philosophy of biology established the broad consensus that there are no strict laws in biology of the sort familiar to physics or even chemistry. Ergo, neither reductionism nor antireductionism about laws is tenable in biology. The entire character of biology as a discipline reflects the considerations that make laws impossible. All of the kinds of biology are functional and therefore have etiologies that reflect natural selection operating on local conditions, and natural selection is constantly changing local conditions. On our planet, natural selection has operated to produce species – spatiotemporally distributed historical individuals whose parts are individual organisms with biological traits that are all functions, or selected effects. This makes biology an essentially historical discipline even when it explains synchronically, as in physiology. Any reformulation of the thesis of reductionism or of antireductionism will have to reflect this fact about the discipline if it is to have a ghost of a chance of illuminating the structure of biology or motivating a biological research program.

Evolution is a mechanism of blind variation and natural selection that can operate everywhere and always throughout the universe. It obtains whenever tokens of matter have become complex enough to foster their own replication and variation so that selection for effects can take hold. Recent experiments in chemical synthesis suggest that this may not be an uncommon phenomenon. Macromolecules are the initial replicators of and also the initial interactors with vehicles (although they are eventually selected for "building" larger interactors or vehicles – chromosomes, cells, tissues, organs, bodies, etc.).

However we express the mechanism of natural selection, its general principles operate exceptionlessly everywhere replicators and their vehicles appear. The principles of the theory of natural selection are the only universally invariant laws in biology.

Beyond the bare theory of natural selection itself, the rest of biology is a set of subdisciplines. Each focuses on a domain historically conditioned by the operation of natural selection on local circumstances during the history of Earth. The functional individuation of biological kinds reflects the vagaries and vicissitudes of natural selection, since biological kinds are the result of selection over hereditary variation among reproducing populations solving (or not) design problems set by the environment. Possible solutions to the same problem are multiple, and one biological system's solution sets a competing biological system's next design problem. Therefore, each system's environment varies over time in a way that turns all putative biological "generalizations" about these systems into historically limited descriptions of local patterns. (This will include even the ones that describe their composition, component constitution, and operation.) Any subdiscipline of biology – from paleontology to developmental biology, population biology to physiology, and including molecular biology – will uncover historically conditioned patterns among lineages owing to the following facts: (1) Its kind vocabulary picks out items generated by a historical process; and (2) its "generalizations" will always be overtaken by evolutionary events. Some of these "generalizations" will describe long-term and widespread historical patterns, such as the ubiquity of nucleic acid as the hereditary material; others of them will be local and transitory, such as the description of the primary sequence of the latest azidothymidine (AZT)-resistant mutation of the AIDS virus.

The generalizations of functional biology are really spatiotemporally restricted statements about the spatiotemporally restricted trends, patterns, and mechanisms that result in them, ones that obtain among and between members of one or more particular species and their parts. Beyond those laws that Darwin uncovered, there are no other nonlocal, ahistorical, counterfactual-supporting (non-*ceteris paribus*) generalizations about biological systems to be uncovered, or at least none to be had that connect kinds under biological – that is, functional – descriptions.

Biological explanation is always at least implicitly historical explanation in which the only biological laws that play a role (along with the laws of physics and chemistry) are the principles of natural selection. It will be important to see why this is true even in reductionism's heartland: molecular biology. Consider, for example, the biologist's explanation of why DNA molecules are composed of thymine while messenger RNA (mRNA), transfer RNA (tRNA), and ribosomal RNA (rRNA) contain uracil. This evidently structural question has a thoroughly historical answer.[4] Long ago on Earth, DNA won the selective race for the best available solution to the problem of high-fidelity information

---

[4] There is a question in chemistry that employs the same form of words: Why does DNA contain thymine when RNA contains uracil? It is answered by a description of the chemical pathway from substrates to the synthesis of the two different nucleic acids, a reductionist's explanation that

storage; meanwhile, RNA was selected for low-cost information transmission and protein synthesis. Uracil is cheaper to synthesize than thymine because thymine has a methyl group that uracil lacks. Cytosine spontaneously deaminates to uracil. DNA with uracil produced by deamination results in a point mutation in the conjugate DNA strand during replication since cytosine pairs with guanine, while uracil and thymine both pair with adenine. A repair mechanism that is evolutionarily available to DNA removes uracil molecules and replaces them with cytosine molecules to prevent this point mutation. The methyl group on thymine molecules in DNA blocks the operation of this repair mechanism when it attempts to remove thymine molecules. Employing thymine, a relatively costly molecule, was a cheaper and/or more attainable adaptation than DNA evolving a repair mechanism that could distinguish uracil molecules that are not the result of cytosine deamination from those that are the result of deamination. So, the DNA repair mechanism was selected for. Meanwhile, the spontaneous deamination of cytosine to uracil in one out of the hundreds or thousands of RNA molecules engaged in protein synthesis will disable only that one molecule, resulting in only a negligible reduction in the production of other copies of the particular type of protein molecule that other tokens of this type of RNA molecule build. Ergo, natural selection for economic RNA transcription resulted in RNA employing uracil instead of thymine. Notice how the explanation works. First, we have two "generalizations": (1) DNA is composed of thymine; and (2) RNA contains uracil. These are not laws, but in fact statements about local conditions on Earth. After all, DNA can be synthesized with uracil in it and RNA can be synthesized with thymine in it. Second, the explanation for the compositions of DNA and RNA appeals to natural selection for solving a design problem set by the environment. Third, tRNA, mRNA, and the various rRNAs are functional kinds, and they have their function as a result of selection over variation. Fourth, we can expect that in nature's relentless search for adaptations and counteradaptations, the retroviruses, in which hereditary information is carried by RNA, may come to have their RNAs composed of thymine instead of uracil if and when it becomes disadvantageous for retroviruses to maximize their rates of mutation. At this point, of course, the original generalizations will, like other descriptions of historical patterns, cease to obtain, but we will have an evolutionary explanation for why they do so, even as we retain our original explanation for why these generalizations about the compositions of DNA and RNA obtain during the period and in the places where they did so. In these respects, explanation in

---

appeals only to the laws of chemistry and physics. Biologists will also be interested in answers to this question, of course.

molecular biology is completely typical of explanation at all higher levels of biological organization. It advances historical explanations in which the principles of the theory of natural selection figure as implicit laws.

So, to be a thesis even worth discussing, reductionism needs to be reconfigured to accommodate itself to historical and historically limited local explanation, for these are the only kinds of explanations that there are in biology, even in molecular biology.

2.1 Reductionism in a Historical Science

Reductionism will thus have to be reformulated as a thesis about (implicitly historical) explanation, not about generalizations or regularities, still less of laws. It will also have to accept that biological explanations are accounts of historical facts, some more widespread and long-lasting than others, but all of them ultimately the contingent results of natural selection operating on boundary conditions. Reductionism needs to claim that the only way to explain one historical fact is by appeal to other more basic historical facts, plus some causal theory that identifies historical facts as relevant causal difference makers. The requirement that explanation identify causes suffices to ensure that there are some laws implicitly or explicitly involved in the explanation. If there are no laws in biology beyond the principles of the theory of natural selection, then the causal explanation of one historical fact by appeal to another will have to appeal, at least eventually and implicitly, to these Darwinian laws, and perhaps even to other laws drawn from physical science, in order to underwrite the causal connections it elucidates. But, by itself, this is hardly a vindication of reductionism. It is little more than the recognition – widely accepted since Hume – that causation always involves the operation of laws at some level or other. Antireductionists can perfectly well accept these conditions on explanation in biology while insisting that biological explanation is fully satisfactory and biologically complete.

There must be more to reductionism that this. Reductionists need to show that the explanations they favor enhance the depth and detail of higher-level biological explanations – improving them and providing something more, something that corrects them, completes them, or fills a lacuna in these explanations, something that makes them *better*.

To see what more there must be to reductionism, recall the distinction between the two different kinds of explanatory tasks in biology: the distinction between proximate and ultimate explanations, which is due to Mayr (1981). Thus, the question as to why butterflies have eyespots may be the request for an adaptationist explanation that accords a function (in camouflage, for instance) to the eyespots on butterfly wings, or it may be the request for an explanation of why at a certain point in development eyespots appear on individual butterfly wings and remain there throughout their individual lives. The former explanation is an ultimate one, while the latter is a proximate one. But notice that the proximate explanation employs functional kinds, selected effects of Darwinian processes of natural selection. So it, too, is implicitly an ultimate explanation, or at any rate presupposes ultimate explanations.

Reductionism must be a thesis about both sorts of explanation. Even to endorse proximate explanations, reductionism needs to hold that ultimate explanations at the level of functional biology are completed, corrected, improved, deepened, and otherwise underwritten by proximate explanations at the molecular level. What kinds of improvements, deepenings, or completions will these reductive explanations provide to the higher-level ultimate explanations?

To expound its thesis about need for the reduction of ultimate to proximate explanations, reductionism adduces another distinction among explanations. It is a distinction that is well-known in the philosophy of history, a division of philosophy whose relevance to biology may now be apparent. The distinction is between what William Dray (1957) called "how-possibly explanations" and "why-necessary explanations." A why-necessary explanation effectively rebuts a presumption that the *explanandum* need not have happened "by showing in the light of certain considerations (perhaps laws as well as facts), it had to happen" (Dray, 1957, p. 161). How-possible explanations show how something could have happened by adducing facts that show that there is after all no good reason for supposing it could not have happened. "The essential feature of explaining how-possibly is . . . that it is given in the face of a certain sort of puzzlement" (p. 165). Dray went on to say, "These two kinds [of explanation] are logically independent in the sense that they have different tasks to perform. They are answers to different questions" (p. 162). But Dray recognized an important asymmetrical relationship between them.

> It may be argued that although, in answer to a "how-possibly" question, all that need be mentioned is the presence of some previously unsuspected necessary condition of what happened . . . nevertheless, this does not amount to a full explanation of what happened. In so far as the explanation stops short of indicating sufficient conditions, it will be said to be . . . an incomplete explanation, which can only be completed by transforming it into an appropriate answer to the corresponding "Why?"
>
> . . . Having given a how-possibly answer it always makes sense to go on to demand a why-necessary one, *whereas this relationship does not hold in the opposite direction.* (p. 168, emphasis added)

Of course, Dray's concern was human history, but the claims carry over into natural history. They suggest an argument that ultimate, how-possibly explanation in functional biology must give way to proximate why-necessary explanation in molecular biology.

Consider the ultimate explanation for eyespots in the butterfly species *Precis coenia*. Notice to begin with that there is no scope for explaining the law that butterflies have eyespots, or patterns that may include eyespots, scalloped color patterns, or edge-bands. There is no such law to be explained. There are, however, historical facts to be explained.

The ultimate explanation has it that eyespots on butterfly and moth wings have been selected for over a long course of evolutionary history. On some butterflies, these eyespots attract the attention and focus the attacks of predators toward parts of the butterfly that are less vulnerable to injury. Such eyespots are more likely to be torn off than more vulnerable parts of the body, and this loss does the moth or butterfly little damage, while allowing it to escape. On other butterflies, and especially moths, wings and eyespots have also been selected for that take the appearance of an owl's head, brows, and eyes. Since the owl is a predator of those birds that consume butterflies and moths, this adaptation provides particularly effective camouflage.

Here, past events help to explain current events via the implicit principles of natural selection. Such ultimate explanations have been famously criticized as "just-so" stories, allegedly too easy to frame and too difficult to test (Gould & Lewontin, 1979). Although its importance has been exaggerated, there is certainly something to this charge. Just because available data or even experience shows that eyespots are widespread does not guarantee that they are adaptive now. Even if they are adaptive now, this is by itself insufficient grounds to claim that they were selected because they were the best available adaptation for camouflage as opposed to being selected for some other function. For that matter, they might not have been selected for at all, but are mere "spandrels," or traits riding piggyback on some other means of predator avoidance or some other adaptive trait.

Reductionists have a reply to this criticism: Adaptationist ultimate explanations of functional traits are "how-possibly" explanations, and the "just-so" story charge laid against ultimate explanation on these grounds mistakes incompleteness (and perhaps fallibility) for untestability. The reductionist has no difficulty with the ultimate, functional, how-possibly explanation, as far as it goes, for its methodological role is to set the research agenda that seeks to provide why-necessary explanations, ones that cash in the promissory notes offered by how-possibly explanations.

How-possibly explanations leave unexplained several biologically pressing issues, ones that are implicit in most well-informed requests for an ultimate explanation. These are the questions of what alternative adaptive strategies were available to various lineages of organisms, and which were not, and the further questions of how the feedback from the adaptedness of functional traits – such as the eyespot – to their greater subsequent representation in descendants was actually effected. The most disturbing lacuna in biology's how-possibly explanations is their silence on the causal details of the feedback loops that operate from the fortuitous adaptedness of traits in one or more distantly past generations to later adaptations, and exactly how and why they ultimately approach any particular locally constrained optimal design. Dissatisfaction with such explanations has been voiced by those suspicious of the theory of natural selection and those amazed by the degree of apparent optimality of natural design, as well as by the religious. The dissatisfaction stems from a widely shared prescientific commitment to complete causal chains, along with the denial of action at a distance and of backward causation. Natural selection at the functional level is silent on the crucial links in the causal chain that convert the appearance of goal-directedness into the reality of efficient causation. The charge that adaptational explanations are unfalsifiable or otherwise scientifically deficient rests in part on their silence about these causal links, and it also rests on the (anti-reductionist) assertion that ultimate explanation is perfectly adequate and has no need to provide the causal links it is silent upon.

Only a macromolecular account of the process of selection for eyespots could provide the details that turn a how-possible explanation into a why-necessary one. Such an account would itself also be an adaptational explanation: It would identify strategies that were historically available for adaptation by identifying the genes (or other macromolecular replicators) that determine the characteristics of the evolutionary ancestors of lepidopterans and that provide the only stock of phenotypes on which selection can operate to move along pathways to alternative predation-avoiding outcomes – leaf color camouflage, spot camouflage, or other forms of Batesian mimicry, repellant taste to predators, Müllerian mimicry of bad-tasting species, etc. The reductionist's "why-necessary explanation" would show how the extended phenotypes of these genes competed and how the genes that generated the eyespot eventually became predominant (i.e., were selected for). In other words, the reductionist holds that: (1) Every functional ultimate explanation is a how-possibly explanation; and (2) there is a genic and biochemical selection process underlying the functional how-possibly explanation. Once completed, reduction turns the merely how-possible scenario of the functional ultimate explanation into a why-necessary proximate explanation of a historical pattern.[5] Note that the reductionist's full explanation is still a historical explanation in which further historical facts – about genes and pathways – are added and are connected together by the same principles of natural selection that are invoked by the ultimate functional how-possibly explanation. But the links in the causal chain of natural selection are filled in to show how past adaptations were available for and shaped into today's functions. All of this will also obtain in every proximate explanation in biology – including molecular biology – that describes functional kinds. They, too, will be implicitly historical.

2.2 Developmental Biology and Explanatory Reduction

Twentieth- and twenty-first-century achievements in developmental biology can illustrate exactly how biology's reductionist research program transforms how-possible explanation sketches into increasingly powerful why-necessary explanations that effect the reductive explanations reductionism foresees. At the same time, these achievements make manifest the need to reconfigure reductionism as a thesis about explanations and not about the deductive derivational relations among laws at various levels of biological organization.

Consider further the developmental biology of butterfly wings and their eyespots. Suppose we observe the development of a particular butterfly wing or, for that, matter the development of the wing in all of the butterflies of the buckeye species, *Precis coenia*. Almost all will show the same sequence of stages, beginning with a wing imaginal disk and eventuating in a wing with such eyespots, and a few will show a sequence eventuating in an abnormal wing or one that is maladapted to the butterfly's environment, owing to the absence of the characteristic eyespots. Rarely, one may show a novel wing or markings that are fortuitously better adapted to the environment than the wings of the vast majority of members of its species. Let's consider only the first case. We notice in one buckeye caterpillar (or in a handful) that during development an eyespot appears on the otherwise unmarked and uniform epithelium of the emerging butterfly wing. If we seek an explanation of the sequence in one butterfly, the general statement that in all members of its species development results in the emergence of an eyespot on this part of the wing is unhelpful. Firstly, this is because examining enough butterflies in the species shows it is false. And secondly, even with an implicit *ceteris paribus* clause, or a probabilistic qualification, we know the "generalization" simply describes a distributed historical fact about some organisms on this planet around the present time and for several

---

[5] In effect, the why-necessary completion of a how-possible explanation provides the details about exactly how the *explanandum* outcome came about. What makes such a how-does-it-come-about explanation distinctively biological is its descriptive role in the explanation of functional and therefore implicitly evolutionary kinds.

million years in both directions. One historical fact cannot by itself explain another, especially not if its existence *entails* the existence of the fact to be explained. That all normal wings develop eyespots does not explain why one does. Most nonmolecular proximate explanations in developmental biology are of this kind; that is, they do little more than summarize sequences of events in the prenatal lives of organisms of a species or, for that matter, in organisms of higher taxa than species.

Here is a typical explanation in twentieth-century developmental biology, which purports to explain the emergence of wings in *Drosophila* (from Wolpert, 1998, p. 320):

> Both leg and wing discs are divided by a compartmental boundary that separates them into anterior and posterior developmental region. In the wing disc, a second compartment boundary between the dorsal and ventral regions develops during the second larval instar. When the wings form at metamorphosis, the future ventral surface folds under the dorsal surface in the distal region to form the double layered insect wing.

This is a purely descriptive account of events in a temporal process recurring in normal *Drosophila* larvae. Its proximate explanation of why a double layer of cells is formed in any one imaginal disk consists in its informing us that that this happens in them all, and that it does so in an order that eventually forms the wing.

But, the reductionist will argue, such proximate descriptions explain nothing. How is the pattern of eyespot development in fact proximally explained? Having identified a series of genes that control wing development in *Drosophila*, biologists then discovered homologies between these genes and genes expressed in butterfly development, and that whereas in the fruit fly they control wing formation, in the butterfly they also control pigmentation. The details are complex, but following out a few of them shows us something important about how proximate why-necessary explanation can cash in the promissory notes of how-possibly explanation and in principle reduce ultimate explanations to proximate ones.

In the fruit fly, the wing imaginal disk is first formed as a result of the expression of the gene *wingless* (so called because its deletion results in no wing imaginal disk and thus no wing) that acts as a position signal to cells, directing specialization into the wing disk structure. Subsequently, the homeotic selector gene *apterous* is switched on only in the dorsal compartment of the imaginal disk, and produces apterous protein, controlling formation of the dorsal (top) side of the wing and also activates two genes, *fringe* and *serrate*, which form the wing margin or edge. These effects were discovered by preventing dorsal expression of *apterous*, which results in the appearance of ventral (bottom) cells on the dorsal wing, with a margin between them and other (non-ectopic) dorsal cells. Still another gene, *distal-less*, establishes the fruit fly's wing tip. Its expression in the center of the (flat) wing imaginal disk specifies the proximo-distal (closer to the body/further from the body) axis of wing development.

It is worth noting the implicit naming convention for many genes in developmental molecular biology: a gene is named for the phenotypic result of its deletion or malfunction. Thus, *wingless* builds wings. Note that genes are individuated functionally and evolutionarily. *Wingless* is so called because of those of its effect that were selected by the environment to provide wings, and similarly for *distal-less*. It's worth recalling that naming genes by their function, even when we could do so by their structure and location, makes imperfect the functional/molecular distinction we are using to contrast higher levels of biological regularity and lower-level molecular ones. Molecular biology is still biology – many of its kinds are functional.

Once the details of the gene control of development were elucidated in *Drosophila*, it became possible to determine the expression of homologous genes in other species, particularly in *Precis coenia*. To begin with, nucleic acid sequencing showed that genes with substantially the same sequences were to be found in both species. In the butterfly, these homologous genes were shown also to organize and to regulate the development of the wing, although in some different ways. For instance, in the fruit fly, *wingless* organizes the pattern of wing margins between the dorsal and ventral surfaces, restricts the expression of *apterous* to dorsal surfaces, and partly controls the proximo-distal axis where *distal-less* is expressed. In the butterfly, *wingless* is expressed in all of the peripheral cells in the imaginal disk that will not become parts of the wing, where it programs their death (Nijhout, 1994, p. 45). *Apterous* controls the development of ventral wing surfaces in both fruit flies and butterflies, but the cells in which it is expressed in the *Drosophila* imaginal disk are opposite those in which the gene is expressed in *Precis* imaginal disks. As Nijhout describes the experimental results:

> The most interesting patterns of expression are those of *Distal-less*. In *Drosophila Distal-less* marks the embryonic premordium of imaginal disks and is also expressed in the portions of the larval disk that will form the most apical [wing-tip] structures ... In *Precis* larval disks, *Distal-less* marks the center of a presumptive eyespot in the wing color pattern. The cells at this center act as inducers or organizers for development of the eyespot: if these cells are killed, no eyespot develops. If they are excised, and transplanted elsewhere on the wing, they induce an eyespot to develop at an ectopic

location around the site of implantation ... [T]he pattern of *Distal-less* expression in *Precis* disks changes dramatically in the course of the last larval instar [stage of development]. It begins as broad wedge shaped patters centered between wing veins. These wedges gradually narrow to lines, and a small circular pattern of expression develops at the apex of each line ... What remains to be explained is why only a single circle of *Distal-less* expression eventually stabilizes on the larval wing disks. (Nijhout, 1994, p. 45)

In effect, the research program in developmental molecular biology aims to identify genes expressed in development and then to undertake experiments – particularly ectopic gene expression experiments – that explain the long-established observational "regularities" reported in traditional developmental biology. The *explanatia* uncovered are always "singular" boundary conditions insofar as the *explananda* are spatiotemporally limited patterns, to which there are always exceptions of many different kinds. The reductionistic program in developmental molecular biology aims first to explain the wider patterns, and then to explain the exceptions – "defects of development" (if they are not already understood from the various ectopic and gene deletion experiments employed to formulate the why-necessary explanation for the major pattern).

The developmental molecular biologist Sean Carroll and his colleagues, who reported the beginnings of the proximal explanation sketched out above, eventually turned their attention to elucidating the ultimate explanation of the eyespots on butterfly wings. These authors wrote:

The eyespots on butterfly wings are a recently derived evolutionary novelty that arose in a subset of the Lepidoptera and play an important role in predator avoidance. The production of the eyespot pattern is controlled by a developmental organizer called the focus, which induces the surrounding cells to synthesize specific pigments. The evolution of the developmental mechanisms that establish the focus was therefore the key to the origin of butterfly eyespots. (Keys et al., 1999, p. 532)

What Carroll's team discovered is that the genes and the entire regulatory pathway that integrates them and that controls the anterior/posterior axis in development in *Drosophila* (or its common ancestors with butterflies) have been recruited and modified to develop the eyespot focus. This discovery of the "facility with which new developmental functions can evolve ... within extant structures" (Keys et al., 1999, p. 534) would have been impossible without the successful why-necessary answer to the proximate question of developmental biology.

Besides the genes noted above, there is another, *hedgehog*, whose expression is of particular importance in the initial divisions of the *Drosophila* wing imaginal disk into anterior and posterior segments. As in the fruit fly, in

Precis the *hedgehog* gene is expressed in all cells of the posterior compartment of the wing, but its rate of expression is even higher in cells that surround the foci of the eyespot. In *Drosophila*, *hedgehog*'s control over anterior/posterior differentiation appears to be the result of a feedback system at the anterior/posterior boundary involving four other gene products, particularly **Engrailed**, which represses another, **Cubitus interruptus** (hereafter "*ci*"), in the fruit fly's posterior compartment. This same feedback loop is to be found in the butterfly wing posterior compartment, except that here the *engrailed* gene's products do not repress *ci* expression in the anterior compartment of the wing. The expression of *engrailed*'s and *ci*'s gene products together results in the development of the focus of the eyespot. One piece of evidence that switching on the *hedgehog–engrailed–ci* gene system produces the eyespot comes from the discovery that in those few butterflies with eyespots in the anterior wing compartment, *engrailed* and *ci* are also expressed in the anterior compartment at the eyespot foci (but not elsewhere in the anterior compartment). Carroll's team conclude: "Thus, the expression of the *hedgehog* signaling pathway and *engrailed* is associated with the development of all eyespot foci and has become independent of the [anterior/posterior] restrictions [that are found in *Drosophila*]" (Keys et al., 1999, p. 534).

Further experiments and comparative analysis enabled Carroll and coworkers to elucidate the causal order of the changes in the *hedgehog* pathway as it shifts from wing production in *Drosophila* (or its ancestor) to producing foci in *Precis coenia* eyespot development. "The similarly between the induction of *engrailed* by *hedgehog* at the [anterior/posterior] boundary [of both fruit fly and butterfly wings, where it produces the intervein tissue in wings] and in eyespot development suggests that during eyespot evolution, the *hedgehog*-dependent regulatory circuit that establishes foci was recruited from the circuit that acts along the anterior/posterior boundary of the wing" (Keys et al., 1999, p. 534).

Of course, the full why-necessary proximate explanation for any particular butterfly's eyespots is not yet in, nor is the full why-necessary proximate explanation for the development of *Drosophila*'s (or its ancestor's) wing. But once they come in, the transformation of the ultimate explanation of why butterflies have eyespots on their wings into a proximate explanation can begin. This fuller explanation will still rely on natural selection, but it will be one in which the alternative available strategies are understood and the constraints specified, in which the time, place, and nature of the mutations narrowed, in which adaptations are unarguably identifiable properties of genes – their immediate or mediate gene products – and in which the feedback loops and causal chains will be fully detailed. As a result, the scope for doubt, skepticism, questions, and methodological critique that ultimate explanations are open to will be much reduced.

2.3 Reductionism's Hostage to Fortune

All biological explanations are at least implicitly ultimate owing to the functional vocabulary they employ to characterize *explanantia* and *explananda*. But if one accepts the arguments and examples that reductionists offer to substantiate their claims about the need to complete, correct, and deepen ultimate explanations offered at the level of functional biology, then reductionism gives one very huge hostage to fortune. Indeed, it gives the antireductionist the basis for a powerful argument that reductionism is a nonstarter even at the level of macromolecules – that even the most complete and adequate explanation at this level, no matter how thoroughly proximate an explanation it may be, will be quite resistant to reduction. And the reason the antireductionist gives is simple and obvious: All explanations in biology, whether proximate or ultimate, invoke – implicitly or explicitly – the operation of the process of natural selection, and this the antireductionist will hold is a patently nonphysical process that operates throughout biology, making each and every explanation proffered in the discipline irreducible to physical explanation.

Reductionism is in need of a convincing argument that, on the contrary, the process that Darwin discovered is a thoroughly physical one whose prominence in biological explanations in no way compromises their reducibility.

3 Reductionism and Natural Selection

The most powerful arguments for physicalist antireductionism turn on the relationship between the theory of natural selection and physical theory. Time and time again, antireductionists have invoked evolutionary facts explainable by natural selection and not explainable by the "gory details" of underlying physical processes as the fundamental barrier to metaphysical reduction. Each time the function of a trait is invoked to explain its presence, character, or mode of action, or even merely to identify it, natural selection is implicitly invoked as the shaper of the trait and as an assurance of the irrelevance of its multiply realized physical implementation. The antireductionist observes that it is these (often implicit) evolutionary facts and regularities that we would miss were we to adopt the point of view of the physicist, the chemist, and even the molecular biologist.

Reductionism needs to show that it is only through the operation of the laws of physics that the process of Darwinian natural selection and all of its adaptational results can emerge. Otherwise, there is space for the possibility that the physical facts have not fixed all of the biological facts, or that some (many, most, or even all) adaptations might be the result of nonphysical processes. Reductionists need to show that the operation of physical law is necessary for the emergence of adaptations. But that's not enough. Reductionism needs to show how the process of natural selection is in fact the result of the operation of physical law alone; that is, it needs to show that physical law is sufficient for the emergence of adaptation by natural selection.

Physicalist reductionism needs an explanation that starts with zero adaptation and builds the rest of the amazing adaptations of biology from the ground up by physics alone. It cannot even leave room for "stupid design," let alone "intelligent design," to creep in. If physicalist reductionism needs a first slight adaptation, it must surrender the claim that the prior physical facts (none of which are constitute adaptations) fix all of the other facts. And it needs to demonstrate that, given the constraints of physics, adaptation could have no other source than natural selection.

The argument for explanatory reduction of the process of natural selection proceeds in two stages. First, it shows that natural selection doesn't need any prior adaptation at all to get started: Beginning with zero adaptations, it can produce all of the rest by physical processes alone. Physics is sufficient for adaptation by natural selection. We need only the second law of thermodynamics to do this. Then with the same starting point – the second law of thermodynamics – we can show that the process Darwin discovered is necessarily the only way adaptations can emerge, persist, and be enhanced in a world where the physical facts fix all of the facts. If the argument works, it will remove the wiggle room for any alternative source of adaptation in the universe. More importantly, it will also accomplish the reduction of natural selection to physics.

3.1 Showing that Physics Is Enough for Adaptation by Natural Selection

The second law of thermodynamics tells us that, with very high probability, entropy – the disorder of things – increases over time.

But the biological realm seems to show the opposite of second-law disorder. It reflects persistent orderliness – start out with some mud and seeds, end up with a garden of beautiful flowers. The ever-increasing adaptation of plants' and animals' traits to local environments looks like the long-term *increase* in order and decrease in entropy. So we have to square the emergence and persistence of adaptation with the second law's requirement that entropy increase.

One might be excused for thinking that if adaptation is orderly and if increases in it are decreases in entropy, then evolution must be impossible.

This line of reasoning makes a slight mistake about entropy and magnifies it into a major mistake about evolution. The second law requires that evolution produce a *net* increase in entropy. Increases in order or in order's persistence are permitted. But these must be paid for by more increases in disorder elsewhere. Any process of the emergence, persistence, or enhancement of adaptation must be accompanied by increases in disorder that are almost always greater than the increases in order. The "almost" is added here because the second law tells us that increases in entropy are only very, very probable, not absolutely invariable. It won't be difficult to show that Darwin's explanation of adaptation – and only Darwin's explanation – can do this.

Natural selection requires three processes: reproduction, variation, and inheritance. It doesn't really matter how any of these three things get done, just so long as each one goes on for long enough to produce adaptations. Reproduction doesn't have to be sexual, asexual, or even easily recognized by us as being reproduction. Any kind of replication is enough. In chemistry, replication occurs whenever a molecule's own chemical structure causes the chemical synthesis of another molecule with the same structure – when it makes copies of itself or helps something else make copies of it. This can and does happen in several different ways, both in the test tube and in nature. The one that is most directly relevant for evolution on Earth is, of course, template matching –the method DNA uses to make copies of itself.

The first step is as follows: When atoms bounce around, some bind to one another strongly or weakly depending on the kind of attraction there is between them – their chemical bond. When the bond is strong, the results are stable molecules. These molecules can only be broken up by forces that are stronger than the bonds. Such forces require more energy than is stored by the stable bond between the atoms in the molecule. Breaking down a stable molecule requires more energy than keeping it together.

Occasionally, these relatively stable molecules can be templates for copies of themselves. Their atoms attract other atoms to themselves and to each other so that the attracted atoms bond together to make another molecule with the same structure. The process is familiar from crystal growth. Start out with a cube of eight atoms in a solution of other atoms of the same kind. They attract another four atoms on each side, and suddenly the molecule is a three-dimensional cross. As it attracts more and more atoms, the crystal grows from a small cube into a large one. The crystal grows in a solution through "thermodynamic noise": the increasingly uneven and disorderly distribution of atoms just randomly bouncing around in the solution as mandated by the second law. The atoms already in the crystal latch on to ones in the solution in the only orientation that is chemically possible, making the regular shape we can see when they get big enough.

A crystal molecule doesn't just have to grow bigger and bigger. Instead, the molecule can set up chemical forces that make two or more other unattached atoms that are just bouncing around bond with one another, making new *copies* of the original crystal. Instead of getting bigger, it fosters the synthesis of more copies of itself.

The process could involve more steps than just simple one-step replication. It could involve intermediate steps between copies. Think of a cookie that is stale enough to be used as a mold to make a cookie cutter that takes the stale cookie's shape. It's a template – to make a new cookie. Make the new cookie, then throw away the cookie cutter, let the new cookie go stale, and use it to make a new cookie cutter. Thus are made lots of copies of the same cookie using one-use cookie cutters. (In a way, this is just how DNA is copied in the cell by using each cookie cutter once, except the cookie cutter is not thrown away, but rather is used in another function – heredity.)

Physical chemists and organic chemists are discovering more and more about how such complicated structures arise among molecules. They are applying that knowledge in nanotechnology – the engineering of individual molecules. Chemists building nanostructures molecule by molecule is remarkable in itself. What is truly amazing is that these nanostructures assemble themselves. In fact, this is the only way nanotechnology works. There are very few molecules that chemists can manipulate one at a time by putting one next to another and then gluing them together. More often, all they can do is set up the right soup of molecules milling around randomly ("thermodynamic noise") and then wait while the desired structures are built by processes that emerge from the operation of the laws of physics. Just by changing the flexibility of small molecules of DNA in the starting soup at the bottom of a test tube and by changing their concentrations, chemists can produce many different three-dimensional objects, including tetrahedrons, dodecahedrons, and buckyballs – soccer ball shapes built out of DNA molecules. Of course, what chemists can do in the lab unguided nature can do better, given world enough and time.

Replication by template matching is even easier than self-assembly, and it works particularly well under conditions that the second law of thermodynamics encourages: The larger the number of molecules and the more randomly the molecules of different kinds are distributed, the better. These conditions increase the chances that each of the different atoms needed by a template will sooner or later bounce into it to help make a copy. In fact, "works well" is an understatement for how completely template replication exploits the second law.

The mixture of atoms bouncing around in a test tube or in nature is very disorderly, and getting more so all the time, as the second law requires. As the disorderly distribution of atoms increases, the chances of different atoms landing on the template also increase. Most of the time, an atom bouncing into a template of other atoms is either too big, too small, or too strongly charged to make a copy molecule that survives. Even if the new atom bonds to the others, the whole copy may break apart due to differences in size or charge or some other chemical property, sending its constituent atoms off to drift around some more, thereby increasing entropy. In most cases – in the lab and out of it – this disorderly outcome of instability in copying is the rule, not the exception. The exception is, of course, a successful duplicated molecule.

The second step required to secure natural selection is to add some variation to the replication. In effect, we are introducing mutation in template copying. At the level of molecules, variation is even easier to introduce than replication. It's imposed on molecules during the process of replication by some further obvious chemical facts working together with the second law of thermodynamics.

One look at the columns of the periodic table of the elements is enough to show how disorder makes chemically similar but slightly different molecules. In the periodic table, fluorine is just above chlorine in the same column. They are in the same column because their atoms react with the atoms of elements in other columns to create new molecules. Chlorine and sodium atoms bond together and make table salt; that means fluorine and sodium atoms will bond together as well (the resulting molecule is a tooth decay preventer). The reason why fluorine and chlorine atoms combine with sodium equally well is that they have the same arrangements of the electrons that do the bonding. All of this means that if chlorine and fluorine molecules are bouncing around and bump into the same template, they may both bond in the same way as other atoms on the template to make similar but slightly different molecules. A template with chlorine molecules in it could easily produce a copy molecule that differs only in having fluorine atoms where chlorine atoms would normally go. *Voilà variation.*

When chemical reactions happen billions of times in small regions of space and time, even a small percentage of exact copies quickly come to number in the millions – as does the percentage of slightly varied copies with one or two atoms of different elements in place of the original atoms. Most of the time, the outcome of this is process is wastage – a molecule that doesn't replicate itself or falls apart, just as the second law requires.

But sometimes – very rarely – variation produces a molecule that is slightly better at replicating, or one that is just a little more stable.

Now we have replication and variation. What about fitness differences, the last of the three requirements for evolution by natural selection? Fitness is easiest to understand at the level of molecules bouncing around in a world controlled by the second law. Molecules that form from atoms are stable for more or less time. Some break apart soon after forming as a result of interference by strong atomic forces such as charge. Some break apart because their bonding is too weak to withstand the force of other atoms that bounce into them or even just pass them by. Some "fragile" molecules will remain intact for a while – they just happen by chance to avoid bouncing into other molecules, ones with stronger charges that pull atoms away from their neighbors. Here, again, the second law rears its head: As molecules bounce around, any amount of order, structure, or pattern almost always gives way to disorder – to entropy. Hardly any molecule is stable for extremely long periods, with the exception of the bonded carbon atoms in a diamond crystal.

There are differences in stability among molecules owing to the variations that inexact replication produces. Differences in stability have an impact on the replication of different types of molecules. A template molecule produces copies simply by random interaction with the atoms that bounce into it or pass by close enough to be attracted. The longer the original templating molecule stays in one piece – that is, the more stable it is – the more copies it can make. Most of its copies will be just as stable as the original template molecule since they will be exact duplicates. They will serve as templates for more copies, and so on, multiplying the copies of the original.

Of course, just as there are differences in the stability of different molecules, there are also differences in their rates of replication. The number of copies of their templates that can be made and their stability will depend on their "environments": on the temperature, the local electric and magnetic fields, and the concentration of other atoms and molecules around them. Consider two molecules that differ from one another along the two dimensions of stability and ease of replication. The first remains intact on average for twice as long as the second; the second templates twice as many copies per unit of time as the first. Over the same period, they will produce exactly the same number of copies. What will the long-run proportions of molecules of the two types be? It will be one to one. As far as producing copies are concerned, the two different molecules will have equal *reproductive fitness*. And, of course, if their differences in stability and replicability don't perfectly balance out, then after a time there are going to be more copies of one type of molecule than of the other.

Molecules randomly bouncing around a region of space and bonding to form larger molecules will eventually randomly result in a few stable and replicating structures. Their structures will vary, and these variations will have effects on the molecules' stability and replication rates. These differences in the ability of molecules to stay intact (stability) and to allow for copies of themselves to form (replicability) will change the proportions of molecules in any region of space. If the chemical and physical conditions in that region remain unchanged for long enough, the ratios of the different types of replicating molecules will eventually settle down to a fixed proportion. At that point, all of the remaining replicating molecules in the region will be equally fit to survive – whether owing to their stability, replicability, or varying combinations of both, they and/ or copies of them persist. In other words, a purely physical process has produced molecular adaptation: the appearance, persistence, and enhancement of molecules with chemical and physical properties that enable them to persist and/or replicate or both. Then at some point the chemical environment changes slightly or greatly: temperatures rise or fall, new and different molecules diffuse through the region, or the magnetic field strengthens or weakens. The process of adaptational evolution starts again, thermodynamically filtering for new stable, replicating molecules adapted to the new conditions.

As this process goes on, two other phenomena become inevitable: The size and the complexity of the replicating molecules will increase. Eventually, there will start to be molecules that enhance each other's stability and/or replication through their chemical relations to each other. There are no limits to the repetition of this process, leading to bigger, more complicated, and more diverse molecules. If conditions are favorable, the result will be really big assemblies of stable and/or replicating molecules, such as RNA and, eventually, DNA sequences and strings of amino acids (i.e., genes and proteins).

The rest is history – that is, natural history. The process we have described begins with zero adaptations and produces the first adaptation by dumb luck, sheer random chance that the second law makes possible. It just had to have happened this way if the physical facts fix all of the facts, including the facts of adaptation.

Molecular biologists don't yet know all of the details of this natural history, or even many of them. Some have been known for a long time. It was in the early 1950s that two scientists – Stanley Miller and Harold Urey – showed how easy it is to obtain proteins, sugars, lipids, and the building blocks of DNA from simple ingredients that would have been available early in Earth's history. All they did was run an electric current through some water, methane, ammonia, and carbon. Chemists have been designing similar experiments ever since, obtaining more and more of the building blocks of terrestrial life. Biologists have discovered that the simplest and oldest of organisms on the planet – the Archaebacteria – probably first emerged at least 2.8 billion years ago, and they still survive in volcanoes at the bottom of the sea. It is there in such volcanoes at the bottom of the ocean – at the highest temperatures and greatest pressures – that one finds chemical reactions spewing out boiling lava and producing the largest quantities of entropy on the planet. This is just what the second law requires to drive thermodynamic noise, and through it to find stable and replicating molecules in a world of random mixing.

How much like the evolution of recognizably biological things – genes, cells, and replicating organisms – is the molecular process we have described? Well, recognizably biological evolution has three distinctive features: First, natural selection usually finds quick and dirty solutions to immediate and pressing environmental challenges. More often than not, these solutions get locked in. Then, when new problems arise, the solutions to old problems constrain and channel the random search for adaptations that deal with newer problems. The results are jury-rigged solutions that are permanently entrenched everywhere in nature.

A second feature of biological evolution is the emergence of complexity and diversity in the structures and behaviors of organisms. Different environments select among variations in different adaptations, producing diversity even among related lineages of organisms. The longer that evolution proceeds, the greater the opportunities for more complicated solutions to continuing design problems.

A third feature of biological evolution is the appearance of cooperative adaptations and competing adaptations. The cooperative adaptations are some-times packaged together in the same organism or sometimes separated in two quite distinct organisms living symbiotically (such as ourselves and the *E. coli* bacteria in our guts). But natural selection also, and more frequently, produces competition between genes, individuals, lineages, and populations within and between species. There is always a chance that "arms races" will break out between evolving lineages of traits in which a new random variation in one trait gets selected for by exploiting the traits of an organism it has been cooperating with, or a new variation randomly appears in a trait that enables it to suddenly break up a competitive stalemate it was locked into.

Each of these three features is found in the nano-evolution of molecules, and each persists as molecular assemblies grow in size, stability, complexity, and diversity to produce organic life.

First, it locks in quick and dirty solutions: Only a few molecular shapes will ever allow for their own self-assembly and copying (by templating or other-wise). These shapes begin the locking in that second-law processes will have to work with in building more stable, more replicable, bigger, and more complicated molecules.

Second, it increases diversity and complexity: Thermodynamic noise constantly makes more and more varied environments with different temperatures, different pH levels, different concentrations of chemicals, different amounts of water, carbon dioxide, or nitrogen, and/or more complicated acids and bases, magnetic fields, and radiation. As a result, there will be a corresponding selection for more and more different molecules. However, they will still be variations on the themes locked in by the earliest stages of molecular evolution.

Third, it fosters cooperation and competition: Some of these molecules will even start to work together due to them – just by luck – having structures that enhance one another's stability or replicability. Others will explode, dissolve, or change their structures when they combine with one another. They will produce new environments that will select for new chemical combinations, some that cooperate and some that compete. This process continues up the ladder of complexity and diversity, eventually producing assemblies of molecules so big that they become recognizable as genes, viruses, organelles, cells, tissues, organs, and organisms, all the way up to our pets, our parasites, our prey, our predators – and us.

To summarize, molecules bouncing against one another inevitably follow a scenario dictated by the second law of thermodynamics. Purely physical and chemical processes in that scenario are *all that is needed* for the emergence, persistence, and enhancement of adaptation through natural selection at the molecular level. Where and when molecules of some minimal size emerge, there will be some natural selection for chemical structures that confer more stability and replicability than other chemical structures confer. These chemical properties are *adaptations*: They have functions for the molecules that exhibit them. They enable the molecules to survive longer than less stable ones in the same chemical bath and to make more copies of themselves than others. As a result of molecular natural selection, these molecules are better adapted to their molecular environments than others.

This is an important outcome for physicalist reductionism. It faces the demand of showing how to get the merest sliver of an adaptation from zero adaptation by purely physical, chemical, thermodynamic processes. More of the same processes can build on that sliver of an adaptation to produce more adaptations and eventually more robust adaptations. But it can't cheat. The process can't just assume the existence of that first sliver. The second law makes the first sliver possible. Variation and selection can take it from there. Next, the reductionist has to show that this is the *only way* adaptations – molecular or otherwise – can emerge: by exploiting the second law operating on zero adaptations to begin with.

3.2 Showing How Physics Makes Natural Selection the Only Way Adaptations Can Arise

The second law makes the merest sliver of an initial adaptation just barely possible, but it makes no guarantees. For all we know, it might happen only once every 13.7 billion years in an entire universe. If the first adaptation survives long enough, the second law allows for improvements, but only if they are rare and energetically costly.

But physicalist reductionism needs to show not just that physics is one possible source of adaptation. Physics won't fix all of the facts unless the second law's way of getting adaptations is the only way to get them. Physicalist reductionism needs to show that blind variation and environmental filtration is the sole route through which life could have emerged in any universe governed by the second law. We have to understand why, in a universe made only by physics, the process that Darwin discovered is the only game in town.

There are three things we need to keep in mind to show this. First, and as already noted, if we are out to explain how any adaptation at all could ever happen, we can't help ourselves to some prior adaptation, no matter how slight, puny, or insignificant. Second, the second law requires that the emergence and persistence of orderliness of any kind be energetically expensive. An explanation of adaptation's emergence and persistence is going to have to show that the process is wasteful. The more wasteful the better, so far as the second law is concerned. Third, physics is going to have to drive the asymmetry of evolutionary change. In the rest of this section, we'll explore the first two of these requirements on the reduction of natural selection to physics. We'll leave the third requirement to the next section.

The very first adaptation has to be a matter of nothing more than dumb luck. It has to be the random result of a very, very mindless process, like a monkey hitting the keys of a typewriter randomly and producing a word. Given world enough and time, the bouncing around of things just fortuitously produces adaptations. The second law insists that initial adaptations – no matter how slight, small, or brief – can't happen very often. And the same has to go for subsequent adaptations that build on these. They will have to be rare and random as well. So, any theory that explains adaptation in general will have to have this feature: that the first adaptation was a fluke, the luck of the draw, just an accident, simply the result of the law of averages. The inevitability of a first, slightest adaptation is ruled out by the second law: To begin with, the second law says that nothing is inevitable, even the heat death of the universe. More importantly, the appearance of the merest sliver of an adaptation is an increase in order, and so at most is improbable.

None of this will be a surprise to Darwinian theory, of course – that's just what the theory says will happen: Variations are random; they are almost always small (in fact, they are almost always molecular); the first ones come out of nowhere, just as a matter of throwing dice or shuffling cards; and mostly they are maladaptive, only rarely being adaptive.

The second law also requires that the process through which later adaptations emerge out of earlier ones be energetically expensive and wasteful: Expensive because building more and more order has to cost more and more disorder, and wasteful because the early steps – the original adaptations on which later ones are built – will be locked in so that less energetically costly ways of building the later adaptations are not available. Every explanation of adaptation will have to share this feature as well. It will have to harness a wasteful process to create order.

' Now, one has only to examine natural selection as it has occurred on Earth to see how expensive and wasteful it is. The combination of blind variation and environmental filtration is great at increasing entropy. In fact, the right way to look at the emergence of adaptation on Earth is to recognize that it is the most wasteful, most energetically expensive, and most entropy-producing process that occurs on the planet. The evolution and maintenance of adaptations by natural selection wins the prize for greatest efficiency in carrying out the second law's mandate to create disorder. Forget design – evolution is a mess. This is a fact about natural selection that is insufficiently realized in biology.

Examples are obvious. A female leopard frog will lay up to 6,000 eggs at a time, each carrying exactly half of all of the order required for an almost perfect duplicate offspring. Yet from those 6,000 eggs, the frog will produce only two surviving offspring on average. Some fish are even more inefficient, laying millions of eggs at one time just to make two more fish. Compared to that, a dandelion is miserly with its seeds. It will spread only 1,000 seeds and produce, on average, one copy of itself. But the human male competes well with profligate fish. He ejaculates millions of sperm, each full of information about how to build a human and almost all capable of fertilizing an egg, and yet 999,999 sperm out of 1,000,000 fail to do so. A high proportion of most organisms go through an entire life cycle of building and maintaining order only to leave no offspring at all. Every one of them has a lot of maiden aunts and bachelor uncles. Insofar as Darwinian processes make reproduction central to adaptation, they rely on a wonderfully wasteful process. It's hard to think of a better way to waste energy than to produce lots of energetically expensive copies of something and then to destroy all of them except for the minimum number of copies needed to do it all over again.

Heredity as practiced on eukaryotes is another amazingly entropy-increasing process. Molecular biologists know that DNA copy fidelity is very high and therefore low in information loss compared to, say, RNA copy fidelity. But think of the costs of this much high fidelity. In every cell there is a vast and complex apparatus whose sole function is to ensure copy fidelity: It cuts out hairpin turns when the very sticky DNA folds back on itself, it proofreads all of the copies, it removes mutations, it breaks up molecules that could cause mutations, etc. In *Homo sapiens*, at least 16 enzymes – polymerases – have so far been discovered whose sole functions are to repair different kinds of error that thermodynamic processes produce in DNA sequences. The costs of high-fidelity heredity – both of building the equipment that protects and corrects the DNA sequences and of operating it – are very great. This is just what the second law requires.

Consider this energy waste on an even larger scale. The evolution of adaptation reflects environmental change over Earth's history. The vast diversification of flora and fauna is also the result of differences between local environments in different places on Earth. From long before continental drift and until long after global warming, environments on Earth have and will change over time and increase entropy as they do so. What is more, once natural selection kicks in, flora and fauna remake their environments in ways that further accelerate entropy increase. When nature started selecting molecules for stability and replicability, it began producing arms races that are more wasteful than anything the Americans and Soviets could ever have dreamed up. From the time natural selection began, there has been a succession of move and countermove in adaptational space made at every level of organization. It has happened within and between every descending lineage of molecules, genes, cells, and organisms. Each line of descent has forever searched through "design space" for ways to exploit its collaborators' and its competitors' adaptations. All of that jockeying for position is wasted when one organism, family, lineage, or species is made extinct by the actions of another organism, family, lineage, or species. This, of course, is what Darwin called the struggle for survival.

Once natural selection hits upon sexual reproduction in the struggle for survival, it spreads so rapidly and becomes so ubiquitous that it's difficult to avoid the conclusion that Darwinian selection must be nature's favorite way of obeying the second law. Natural selection invests energy in the cumulative packaging of co-adapted traits in individual organisms just in order to break them apart in meiosis – the division of the sex cells. Then it extinguishes the traits and their carriers in a persistent spiral of creative destruction. Consider that 99 percent of the species that have been extant on this planet are now extinct. That is a lot of order relentlessly turned into entropy!

It's well-known that every major change and many minor ones in the environment condemn hitherto fit creatures to death and their lineages to extinction. As environments change, yesterday's adaptation becomes tomorrow's maladaptation. In fact, it looks like three different cataclysmic events have repeatedly killed off most of the life-forms on Earth. The dinosaur extinction of 65 million years ago, owing to the combination of an asteroid collision on the Yucatan peninsula and widespread volcanic events elsewhere, is well established. There are no dinosaur bones in any of the younger layers of stone around the world, but there is a layer of iridium – an element found in much higher concentrations in asteroids than on Earth – spread evenly around vast parts of Earth centered on the Yucatan in the layers of rock that are 65 million years old. In that layer, the iridium is 1,000 times more concentrated than elsewhere above or below it in Earth's crust. At a stroke, or at least over only a few years, all of the vast numbers of dinosaur species, which had been around and adapting to their environment beautifully for 225 million years, just dis-appeared. That's what made the world safe for the small, shrew-like mammals from which we are descended. The fossil record reveals a bigger extinction event 500 million years ago on Earth, and an even more massive extinction after that, 225 million years ago: the Permian–Triassic extinction in which three-quarters of all ocean-living genera and almost 100 percent of ocean-dwelling species, along with 75 percent of land species, became extinct. This is order-destroying waste on a world historical scale.

Long before all of this, it was the buildup of oxygen in the oceans and the atmosphere that killed off almost everything on Earth. But how could oxygen be a poison? Remember, yesterday's adaptation can be tomorrow's maladap-tation. Life started in the oceans with anaerobic bacteria – ones that don't need oxygen. In fact, they produce oxygen as a waste product in the same way we produce carbon dioxide. Just as plants clean up our mess by converting carbon dioxide into oxygen and water, the environment cleaned up all of that oxygen pollution through molecular action, binding the oxygen to iron and other metals. At some point, the amount of oxygen waste produced by the anaerobic bacteria exceeded the absorption capacity of the environment. As a result, they all began to be poisoned by the increasing levels of oxygen around them. Around 2.4 billion years ago, these bacteria were almost all completely wiped out, thereby making enough space for the aerobic bacteria – the ones that live on oxygen and produce carbon dioxide as a waste product. We evolved from these latter bacteria.

Can any other process produce entropy as fast and on such a scale as natural selection? Just try to think of a better way of wasting energy than this: Build a lot of complicated devices out of simpler things and then destroy all of them except the few you need to build more such devices. Leaving everything alone and not building anything won't increase entropy nearly as fast. Building very stable things, such as diamond crystals, will really slow it down, but building adaptations will use up prodigious amounts of energy. Adaptations are complicated devices: They don't fall apart spontaneously – they repair themselves when they break down. They persistently get more complicated and so use even more energy to build and maintain themselves. Any long-term increase in the number of adapted devices without increased energy consumption would make a mockery of the second law. If such devices are ever to appear, besides being rare, they had better not persist and multiply, unless by doing so they inevitably generate more energy wastage than there would have been without them. This is the very process Darwin envisioned: in Tennyson's words, "Nature, red in tooth and claw."

The second law allows adaptations, but only on the condition that their appearance increases entropy. Any process competing with natural selection as the source of adaptations has to produce adaptations from non-adaptations, and every one of the adaptations it produces will have to be rare, expensive, and wasteful. This requirement – that building and maintaining orderliness always has to cost more than it saves – rules out all of natural selection's competitors as the source of adaptation, diversity, and complexity.

Could there be a process that produces adaptations that is less wasteful than the particular way in which Darwinian natural selection occurred on Earth? Probably. How wasteful any process producing adaptation can be depends on the starting materials and on how much variation emerges in the adaptations built from them. But every one of these possible processes has to rely on dumb luck to produce the first sliver of an adaptation. In that respect, they would still just be more instances of the general process Darwin discovered: blind variation and environmental filtration. A process that explained every later adaptation by more dumb luck shuffling and filtering the earlier adaptations would still be Darwinian natural selection. It would be Darwinian natural selection even if the process was so quick and so efficient as to suggest that the deck was stacked. So long as the deck wasn't stacked to produce some prearranged outcome, it would still just be blind variation and environmental filtration. Any deck stacking – a process of adaptational evolution that started with some unexplained adaptation already in the cards – is ruled out by physics.

3.3 Physical Asymmetry and Adaptational Evolution

Only three things are required to show that, given the laws of physics, the only way that adaptations could have emerged is by natural selection. The first is the requirement that the first and all subsequent adaptations be random and rare events, and the second is that the process by which adaptations persist and improve be energetically expensive. But the third requirement we need to show that natural selection is the only game in town for building adaptations is a rather deeper and less widely noticed feature of the second law.

No matter what brings it about, the process of adaptation is different from the more basic physical and chemical processes in nature. The latter are all "time symmetrical" – adaptation is not. But the only way a time-asymmetrical process can happen is by harnessing the second law.

A time-symmetrical process is one that is physically reversible. One example of such a time-reversible process is well-known: Any set of ricochets on a billiards table can be reproduced in exactly the opposite order. Here are some more examples: Hydrogen and oxygen can combine to produce water, but water can also release hydrogen and oxygen. Even the spreading circular waves made when a drop of liquid falls into a pool can be reversed to move inward and expel the drop upward from the surface. No matter in what order the basic chemical and physical processes go, they can go in the reverse order as well.

The second law creates all asymmetrical processes and gives them their *direction in time*. Now, the evolution of adaptations is a thoroughly *asymmetrical* process. Take a time-lapse film of a standard natural selection experiment. Grow bacteria in a Petri dish. Drop some antibiotic material into the dish. Watch the bacterial slime shrink until a certain point when it starts growing again as the antibiotic-resistant strains of the bacteria are selected for.

Now try reversing the time-lapse video of the process of bacterial selection for resistance. What you will see just can't happen. You will watch the population of the most resistant strain diminish during the time in which the antibiotic is present. After a certain point you will see the spread of the bacteria that can't resist the antibiotic, until the drops of the antibiotic leave the Petri dish altogether. But that sequence is impossible. It's the evolution of maladaptation – the emergence, survival, and spread of the less fit.

There is only one physical process available to drive asymmetrical adaptational evolution: the entropy increase required by the second law of thermodynamics. Therefore, the second law must be the driving force in adaptational evolution. Every process of adaptational evolution – whether it's the one

Darwin discovered, or any other – has to be based on second-law entropy increase.

The physical facts – the starting conditions at the big bang, plus the laws of physics – fix all of the other facts, including the chemical and biological ones. All of the laws of physics except the second law work backward and forward. So, every one-way process in the universe must be driven by the second law. That includes the expansion of the universe, the buildup of the chemical elements, the agglomeration of matter into dust motes, the appearance of stars, galaxies, solar systems, and planets, and all other one-way processes. And that will eventually include, on one or more of these planets, the emergence of things with even the slightest, merest sliver of an adaptation. We can put it even more simply. In a universe of things moving around and interacting on paths that could go in either direction, the only way any *one-way* patterns can emerge is by chance, here and there, when conditions just happen to be uneven enough to give the patterns a start.

That these rare *one-way* patterns will eventually just peter out into nothing is a consequence of the second law. Consider the one-way process that built our solar system and maintains it. It may last for several billion years, but eventually the pattern will be destroyed by asteroids or comets or the explosion of the sun or the merging of the Milky Way with other galaxies, whichever comes first. That's entropy increase in action on a cosmic scale. On the local scale, entropy increase will occasionally and randomly result in adaptational evolution, but it will eventually be effaced. That is the fate of all adaptations that emerge in a universe where all of the facts are fixed by the physical facts.

Moreover, because entropy increase is a one-way street, the second law is also going to prevent any adaptation-building process from retracing its steps and finding a better way to skin the cat. Once a local adaption appears, it can't be taken apart and put together in different, more efficient, less entropy-increasing ways. The only way to do that is to start over independently. Adaptation building has to produce local equilibria in stability and replication that get locked in, built into the hereditary information that carries the adaptation. These local equilibria have to be worked around in the creation of new adaptations. Natural selection is famous for producing such examples of inferior design, Rube Goldberg complexity, and traits that could only have been "grandfathered in." It's not just the oft-cited example of the blind spot in the mammalian eye resulting from the optic nerve's coming right through the retina. An even more obvious case is the crossing of the trachea and the digestive system at the pharynx. We think of the giraffe's neck as an adaptation par excellence. But the nerve that travels from a giraffe's brain to its larynx ended up going all of the way down the neck, under the aorta, and back up – a 20-foot detour it didn't need to take, but that got locked in before neck length began to be selected for.

Any adaptation-creating process has to produce suboptimal alternatives all of the time. It has to do this not just to ensure entropy increase, but also to honor the one-way direction that the second law insists on. Perhaps our most powerful adaptation is the fact that our brains are very large. They have enabled us to get to the top of the carnivorous food chain everywhere on Earth. But this is only the result of a piece of atrocious design. The mammalian birth canal is too narrow to allow large-brained babies to pass through. This bit of locked-in bad design meant that the emergence of human intelligence had to await random changes that made the skull small and flexible at birth and thus delayed brain growth until after birth. This is where the large fontanel – the space separating the three parts of the infant's skull – comes in. Once it emerged, the infant had room to get through the birth canal with a skull that would expand afterward to allow a big brain to grow inside it. But brain growth after birth introduced another huge problem: the longest period of childhood dependence of any species on the planet. All of this maladaptation, corner-cutting, and jury-rigging is required by the second law of any process that produces adaptations.

What all of this comes to is a vindication of reductionism as a research program that can invoke natural selection without any inconsistency – that can identify it as a purely physical process, the only one that, consistent with the second law of thermodynamics, could even bring adaptations into existence, improve them, and ensure their persistence, spread, and eventual maladaptation and extinction. The reality that the process Darwin discovered is a physical one, mandated by physics alone, enables the reductionist – at least in principle – to convert every implicit and explicit ultimate explanation of biology into a proximate one in which the links of the causal explanatory chain are forged by this physical process. When natural selection imposes a local regularity in the biological domain, including ones that obtain long enough to provide some explanations, it in effect gives the regularity a physical basis and opens up a route to its explanatory reduction. It motivates the research program that reductionism long advocated – and that won a much larger adherence among philosophers once it was rebranded as "mechanism." In the next section, we explore exactly how and why this happened.

4 Reductionism Makes Way for Mechanism

In the period after Watson and Crick's breakthrough ("1953 and all that"), a research program calling itself reductionism burgeoned in the life sciences, especially in developmental biology and neuroscience, employing technological innovations – many exploiting the power of regulatory somatic genes – whose mode of action was itself a reductionist triumph.

Meanwhile, reductionism remained an unpopular doctrine in the philosophy of biology, or rather the word continued to have a pejorative connotation among philosophers of science. Perhaps this is because, as noted in Footnote 1, it has been used as a term of abuse in the wider academic and intellectual culture. Section 1 mentioned how easy it is to associate the adjective "reductive" with the noun "reduction." The former is an epithet employed to criticize a claim, doctrine, or theory for ignoring causally relevant variables in a domain or for denying their role. For example, one often finds the label "reductive" rightly imposed on claims that biological traits are the result of nature not nurture, of the action of genes, or of macromolecules alone, and not of the environmental factors that interact with them. It is plain that reductionism as a claim about how explanations should be improved or research programs advanced has nothing to do with "reductive" doctrines of any sort. It never countenances ignoring causally relevant variables. It insists that such variables be provided with a physical mechanism of operation. The word "mechanism" here is significant.

But "reduction" and "reductionism" have never really been able to shake this association with a clearly unwarrantable approach to explanation and theorizing. As a result, by the beginning of the twenty-first century, the terms had begun to be supplanted as labels by the term "mechanism", which is employed to describe the demand that biology uncover mechanisms, and ultimately molecular, physical mechanisms, and that biological systems and processes be explained by identifying the mechanisms that compose them. In the words of two influential exponents of the view, mechanism "is a structure performing a function in virtue of its component parts, component operations, and their organization. The orchestrated functioning of the mechanism is responsible for one or more phenomena" (Bechtel & Abrahamsen, 2005, p. 423). Because the mechanism is causally responsible for the phenomenon, mechanists hold, it provides the best and most correct, adequate, or complete explanation of the phenomenon.

In making this claim, mechanists have been encouraged by the trend of developments elsewhere in the philosophy of science, particularly in the analysis of the nature of explanation. Shortly before mechanism began to burgeon as a thesis about the best way to explain phenomena, a quasi-consensus had been forming around the claim that explanation was not fundamentally nomological (Nagel, 1961), or a matter of unification (Kitcher, 1995) or even the provision of erotetically (pragmatically) successful answers to agents' why questions (van Fraassen, 1985). Instead, following Salmon (1984), there was increasing agreement that scientific explanations cite causes, and in particular causally necessary "difference makers." The *locus classicus* of this approach was provided by Woodward (2005), whose work focused on how experimentally to identify causal difference makers, at least in the ideal case. The details of his approach were music to the ears of mechanists for reasons that will become apparent in this section.

4.1 How Mechanisms Explain

The picture shown in Figure 1, which is endorsed by many mechanists, communicates the way mechanisms explain.

Mechanism, as we will see, is very close to reductionism, perhaps even being an old wine in a newer bottle. Its advocates hold that "biology has become the search for mechanisms ... Biologists look for mechanisms because they serve the three central aims of science: prediction, explanation, and control" (Craver & Darden, 2013, p. 6). Mechanism is the claim that

> Across the life sciences the goal is to open black boxes and to learn through experiment and observation which entities and activities are components in a mechanism and how those components are organized together to do something that none of them does in isolation ... One cannot understand biology ... without understanding ... how mechanism schemas are constructed, evaluated, and revised. (Craver & Darden, 2013, pp. 3, 10)

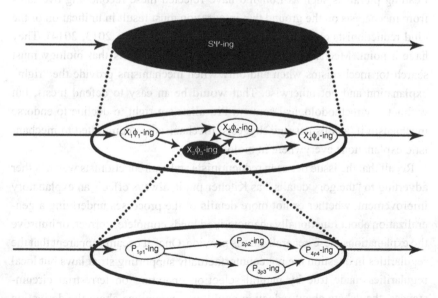

Figure 1 S does ψ owing to its components X_i doing ϕ, each X_i does ϕ owing to its components P_i doing ρ, and so on.

(Credit: Pamela Speh, used with permission.)

What really gives mechanism its reductionist pedigree is its inheritance of post-positivism's embrace of the unity of science, a unity won by the uncovering of inter-field (Darden, 2006) mechanisms everywhere:

> The integration of biology is forged by building mechanism schemas that span many different levels, bridge across many different time scales, and that satisfy evidential constraints from many areas of biology (chemistry and physics too). From the perspective of a given phenomenon, one can *look down* to the entities and activities composing it. One can *look up* to the higher-level mechanisms of which it is a component. One can *look back* to the mechanisms that come before it or by which it developed. One can *look forward* to what comes after it. One can *look around* to see the even wider context with which it operates. The adequate explanation of many biological phenomena requires describing a temporally extended and multilevel mechanism. This is why many fields, working at multiple levels, often must integrate their work in the discovery of mechanisms. (Craver & Darden, 2013, p. 163)

This is a picture of mechanism all the way down, all the way across, and all the way back.

Despite this commitment to causal mechanistic unification, some mechanists have expressed sympathy for pluralism (see Section 1), or at least the hope that their views will not be treated unsympathetically by pluralists (Tabery, 2014). Leading pluralists such as Longino have rejected these reconciling overtures from mechanists on the ground that integration must result in unification of the kind reductionists embrace and pluralists reject (Longino, 2013, 2014). They have a point. Mechanism can hardly be the trivial thesis that biology must search for mechanisms when and only when mechanisms provide the "right" explanation and not otherwise. That would be an easy to defend thesis, but without any methodological content. Pluralists are right to decline to endorse mechanism if, as this passage suggests, mechanists are committed to mechanistic explanations everywhere in biology.

Recall that the issue between reductionists and antireductionists was whether adverting to "the gory details," as Kitcher put it, always effects an explanatory improvement, whether or not more details of the processes underlying a generalization about functionally characterized kinds complete, correct, or improve the explanations these generalizations provide. Once it became apparent that the regularities in question are not counterfactually supporting strict laws but local regularities made true by natural selection operating on terrestrial circumstances, the debate about reduction could not remain one about the derivation or non-derivation of laws. It had to become one about the explanation of particular spatiotemporally restricted biological patterns, finite sets of individual events, states, and processes. Mechanists accept that the *explanantia* in the life sciences are not strict laws, but they sometimes describe the behavior of particular systems using a "phenomenal(ly adequate) model." This kind of model correctly describes the quantitative relations among the variables that characterize the behavior of several different systems and so purports to explain their behaviors without going into the gory details of how each of them does it.[6] Like reductionists, mechanists hold that such models do not adequately explain the behaviors they systematize (e.g., see Kaplan, 2011). Without a specification of the causal mechanism that realizes the often purely mathematical relation that the phenomenal model describes, the model cannot discriminate inputs from outputs, joint effects from causal chains, or *explananda* from *explanans*. Mathematical models, even when empirically adequate to phenomena, lack the asymmetry that causal explanations must report.

Like antireductionists, their opponents insist that often such phenomenal models are explanatorily adequate and do not require supplementary or supplanting detail. Echoing a complaint made by Kitcher and other antireductionists, these philosophers deny that more accurate detail concerning mechanisms always improves such models and the explanations they purport to describe (Batterman & Rice, 2014, p. 352).[7] Many critics (and even some mechanists) emphasize that "abstractions" from a mechanistic explanations can convey understanding: "[A]bstract models play a role in explaining the particular behavior of particular systems" (Bechtel & Levy &, 2013, p. 259). These philosophers deplore "a gap in the mechanistic outlook itself, in which room has not been made for the explanatory fruits of abstracting away from structural detail" (Levy, 2013, p. 488). But the argument for the explanatory power of abstraction from mechanist details is a friendly objection compared to those that hold that there are explanatory forces in the biological domain that are not mere abstractions from more generic descriptions of the same causal variables. Abstraction is a matter of describing the same items in less detail. All too often in the life sciences, nonmechanistic explanations trade in completely different forces and factors, not merely in more abstract versions of the ones mechanists (and before them reductionists) countenance.

This is the point of those who contrast ultimate evolutionary explanations with proximate mechanistic ones. The former are not just more abstract versions of the latter in their view. We return to this matter in Section 4.2.

---

[6] Cf. Piccinini and Craver (2011) and Craver and Kaplan (2011), who argue that these dependencies are not explanatory at all.

[7] Sometimes they argue for this conclusion from human cognitive and computation limitations and scientists' interests (e.g., Chirimuuta, 2014, p. 149). These considerations cannot be metaphysically relevant.

At least some mechanists are prepared to embrace at least a qualified version of the reductionist claim that more detail – and in particular more detail about constituent parts and their role in bringing the *explananda* about – always enhances explanation. Craver and Kaplan (2020) articulate such a position. Assuming that explanations are causal in the sense articulated first by Wesley Salmon and later by James Woodward, they introduce the notion of *Salmon-completeness*:

> SC: The Salmon-complete constitutive mechanism for P vs. P' is the set of all and only the factors constitutively relevant to P vs. P'.[8]

They plug SC into a version of the thesis that, suitably relativized, More Details Are Better (MDBr)

> MDBr: If model M contains more explanatorily relevant details than M* about the SC mechanism for P vs. P', then M has more explanatory force than M* for P vs. P', all things equal.

Then, recognizing that models may not be the only vehicles of explanatory theorizing in science, they broaden this "More Details Are Better" thesis in a variety of ways that strengthen mechanism's reductionist pedigree.

4.2 Mechanism and Higher-Level Explanation

Like reductionists before them, mechanists want to accord some explanatory adequacy to higher-level regularities, generalizations, and the causal dependencies they report. No more than reductionists do mechanists wish to be eliminativists. They need to assert MDBr while accepting that some higher-level causal models have explanatory force even as these models are altogether silent about any SC.

To accord causal explanatory status to such ("phenomenological") models, mechanists need to show how they and the explanations they provide have some real explanatory force independent of a specification of the mechanisms that underlie them. One way to do this might be to insist that high-level causation is autonomous from lower-level causation. If we could show this to be consistent with MDBr, the mechanist would be able to honor the explanatory power of higher-level models by holding that these models identify distinct higher-level causes. There is a danger in this strategy for mechanism. Once distinct higher-level causes are acknowledged, some philosophers and opponents of reductionism throughout the sciences may argue that these causes can operate in

---

[8] P vs. P' reflects their exploitation of a contrastivist approach to explanation, which is invoked to avoid some well-known but irrelevant counterexamples to various causal theories of explanation.

downward directions, exercising effects on lower levels of organization, something no mechanist, reductionist, or physicalist should want to countenance.

Some mechanists have made proposals that could thread this needle: reconciling autonomous higher-level causation with requirements such as MDBr without opening the floodgates to downward causation. Writing with Bechtel, another strong advocate of mechanism, Craver has advanced such an argument seeking to show that mechanism is compatible with higher-level causation, but not downward (or upward) causation (Craver & Bechtel, 2007). If this is right, then dependencies that obtain at higher levels may play a role in at least some explanations that are adequate at their own levels. In that case, explanations honoring MDBr's requirements could be treated as improvements, deepening explanations that are already adequate (though improvable) by the standards of the special sciences that formulate them.

What is more, if MDBr can be satisfied by explanations that don't involve upward causation from lower levels to produce higher-level causation, it will remain open for mechanists to hold that higher-level causal sequences really are autonomous from lower-level causal sequences. This would be the best assurance of the autonomy of explanations that report higher-level causal sequences from deeper explanations that appeal to the mechanisms that bring about the higher-level regularities.

Bechtel and Craver in fact seek to do this. Assuming that higher-level causation obtains, they have claimed that mechanism rules out both downward causation from wholes to parts and upward causation from parts to wholes (Craver & Bechtel, 2007, p. 548; Bechtel, 2008). Instead, what appears to be "inter-level" (upward or downward) causation is always just a combination of "constitution" plus *intra*-level (same level) causation. Thus, high-level causation is preserved independent of lower-level causation and provides a basis for autonomous higher-level explanations. Demands such as MDBr are preserved since the higher-level causation, although independent of lower-level causation, consists in lower-level causation plus constitution of higher-level objects and processes by their mechanical components suitably arranged, and such explanations can be improved by greater detail. Craver and Bechtel (2007) write: "We assume that there are higher-level causes and, further, that all higher-level causes are fully explained by constitutive mechanisms" (p. 548). So, distinct higher-level causes exist, but they are fully explained in ways that honor MDBr.

Craver and Bechtel's proposal raises a question about in exactly what "constitution" consists in terms of how it differs from causation and what special contribution it makes, a contribution that makes higher-level causation real without being mysteriously "emergent" – distinct, autonomous, and capable of downward effects. Elsewhere, Craver (2007) seeks to deliver a conception of "constitution" that does these things. Mechanism requires that particular mechanisms constitute the phenomena they explain. In Figure 1, this is the requirement that the higher-level phenomenon S's ψ-ing is identical to the mid-level mechanism composed of all the X_i's ϕ-ing in a particular causal network. (Each of the X_i's ϕ-ing is a phenomenon constituted by each of its components X_i ρ-ing.) None of the X_i's ϕ-ing can cause S's simultaneous ψ-ing, as they occur at the same time in a subregion of S (causation is assumed to be temporally asymmetrical in the life sciences). But each of the X_i's ϕ-ing is *constitutively relevant* to S's ψ-ing if and only if X_i's ϕ-ing is a part of S's ψ-ing, and (roughly) changes in the state of S – the phenomenon – cause (slightly later) changes in the state of an X_i, a component of the mechanism that constitutes it, while changes in the state of the X_i cause (slightly later) changes in the state of S. Mutually, the effects of component changes to changes in the higher-level phenomena and vice versa are what the constitutional (as opposed to the causal) relevance of component processes to the whole phenomenon consists in. The set of components of a mechanism, each constitutively relevant to a higher-level phenomenon itself, constitutes that phenomenon.

This account of how mechanisms constitute phenomena makes higher-level causation non-mysteriously distinct from lower-level causation, as it also requires constitutive relevance, which is itself a matter of lower-level causation. (A change in S's ψ-ing can only occur through some "intervention" on some component part of the mechanism that constitutes it, presumably a different part from the X_i in question when the relation of constitutive relevance is established by experiment.)

4.3 Mechanism and Functional Individuation in the Life Sciences

Can we reconcile MDBr with the explanatory adequacy of higher-level expla-nations regarding their appeal to functions, functionally characterized systems, and the *ceteris paribus* regularities that advert to them? This is the main challenge facing mechanism as the inheritor of reductionism. Mechanists gen-erally recognize the role of functional kinds in the life sciences and their importance as starting points for mechanistic analysis. Thus, Piccinini and Craver insist that "Functional properties are an undetachable aspect of mechan-istic explanations. Any given explanatory text might accentuate the functional properties at the expense of the structural properties, but this is a difference of emphasis rather than difference in kind. The target of the description is in each case a mechanism" (Piccinini & Craver, 2011, p. 17). So, mechanism requires functional individuation to get started: Mechanisms are always mechanisms for

X, whether X is what they do or what they attain, and where X is always a function – a biologically salient selected effect. But mechanists don't take functions seriously as autonomous irreducible properties. They can't – they are committed to treating functional descriptions as descriptions of physical properties and mechanisms. But their dependency on functional description is fraught with difficulties.

As noted earlier in this section, one way in which the explanation of a higher-level dependency can satisfy MDBr is when the kinds that figure in the dependency differ from those in the mechanistic explanation by simply being more abstract kinds from which (irrelevant?) details of physical implementation have been eliminated. The abstract/concrete distinction is an obvious one. For example, "force" is a more abstract kind than "gravitational force" or "electro-motive force." "Acid" is a more abstract kind than "sulfuric acid." Newton's second law obtains for all forces – it does not specify what kinds of forces produce accelerations. Similarly for mechanism – assimilating functional kinds to MDBr will not be challenging if the functional kinds of biology differ from the physical kinds of their mechanistic explanations only along the dimension of abstractness versus concreteness in detail. Under what conditions will this relationship between kinds obtain in the life sciences?

This is where multiple realizability comes into the picture for mechanism. For example, as noted in Section 1, the wing evolved independently 40 times in evolutionary history. If there were interesting dependencies between flight and wings, demanding their explanation satisfy MDBr would be hopeless, since differences in the mechanisms that explain flight don't differ merely in the details of implementation – they differ radically. Consider bird wings, insect wings, pterosaur wings, and bat wings. The mechanistic explanation of how wings do their job will be so highly disjunctive as to lack any explanatory unity whatsoever, and this disjunction of mechanisms will not preserve any explanation that invokes the general properties that wings share, *ceteris paribus*.

Because natural selection is blind to differences in structure that have common effects for survival and reproduction, all of the functional kinds that it has produced in the biological domain will have multiple realizations. Just as these multiple realizations confronted reductionism with difficulties, they threaten to bedevil mechanism as well.

The concrete details that MDBr demands about functionally described phenomena will have almost always to be disjunctive, sometimes extremely disjunctive. And the disjunctive details that MDBr demands be adduced may not always improve on explanations couched in functional terms that are not just abstractions from more concrete mechanistic descriptions.

Consider, for example, Fisher's well-known model of the 50:50 sex ratio in sexually reproducing species. Fisher's model explains this ratio without any reference to the mechanism of reproduction shared by all of these species. Fisher wrote down a proof from some harmless idealizations that when one sex or the other exceeds 50 percent of the population, natural selection confers higher fitness on females that disproportionately bear offspring of the minority sex. This process will shift the population back to 50:50, and when it over-shoots, it will enhance the fitness of females that disproportionately give birth to offspring of the minority sex so that, other things being equal, *ceteris paribus*, the sex ratio always remains close to 50:50. The *ceteris paribus* clause is important as there are species, such as humans, that don't maintain this ratio in live births, but have a slightly higher male-to-female ratio (the exception being explained by another application of Darwinian considerations), and paper wasps that violate the ratio radically (again as exceptions that prove the rule, as they result from Darwinian selection operating on special environments). The MDBr thesis has to find a way to preserve the explanatory goodness of this model while showing that mechanistic details improve on the model's expla-natory value for the vast number of cases where it works.

Excluding cases of ultimate explanation from the writ of MDBr may be criticized as ad hoc. Given the importance of such explanations in the life sciences, such an exclusion would reduce the significance of mechanism as a demand on explanation. One challenge to the agenda of mechanism, then, is to show that natural selection is a mechanism, one that can be described at various levels of abstraction – that models such as Fisher's are at the higher levels and that the causal mechanisms that underwrite them fill out the concrete details of particular cases in ways that vindicate MDBr and the original abstractions that later details improve (see Skipper & Millstein, 2005; Havstad, 2011). This problem also challenged reductionism, of course, and mechanists may want to avail themselves of the strategy of invoking the second law of thermodynamics to underwrite natural selection as a mechanism (sketched out in Section 3) to solve it.

But there may be harder cases for mechanists to reconcile. Consider Kitcher's explanation of Mendel's law of independent assortment discussed in Section 2. The regularity is given by:

> (G) Genes on different chromosomes, or sufficiently far apart on the same chromosome, assort independently.

Note that (G) already includes mechanistic details (genes are located on chro-mosomes) that Mendel's original regularity lacked.

Kitcher's non-reductive explanation of (G) was given in the following principle:

> (PS) There are some basic *entities* that come in pairs. For each pair, there is a correspondence relation between the parts of one member of the pair and the parts of the other member. At the first stage of the process, the entities are placed in an *arena*. While they are in the arena, they can exchange segments, so that the parts of one member of a pair are replaced by the corresponding parts of the other members, and conversely. After exactly one round of exchanges, one and only one member of each pair is drawn from the arena and placed in the *winners box*.
>
> In any PS-process, the chances that small segments that belong to members of different pairs or that are *sufficiently far apart* on members of the same pair will be found in the winners box are independent of one another. (G) holds because the distribution of chromosomes to gametes at meiosis is a PS-process. (Kitcher, 1984, pp. 341–343, emphasis added)

This, Kitcher claimed, was a full explanation of (G), even as it prescinds from any mention of genes or their mechanism of action.

Mechanists will be pleased to note that (PS) in fact provides a mechanism that explains (G). But as Figure 1 makes clear, they will also have to explain how the mid-range claims about *entities, arenas, sufficiently far apart,* and *winners' boxes* that the mechanism of (PS) consists in can be improved by deeper mechanistic details about how far apart is sufficiently far apart for entities, arenas, and winners' boxes to work as required to generate (G). The same will go for other functional kinds carved out by natural selection.

Mechanists have sought to mitigate these problems shared with reductionism in a few different ways. For one thing, mechanists recognize that all biological explanation is implicitly the explanation of particular facts, events, states, processes, and local dependencies. Mechanism is token explanation, not type explanation. Accordingly, it does not need to formulate nomological bridge principles – type identities – that enable one set of laws about broader natural kinds to explain laws about narrower natural kinds. Token identities between (finite sets of) functionally characterized items and the mechanisms that constitute them will suffice for its explanations. When these sets are small, the scope for radical differences in mechanistic realizations of a single function is reduced (Bechtel, 2009). Moreover, many of the functional kinds of organismic biology, especially botanical, anatomical, and behavioral kinds, themselves give way on further analysis to finer functional categorization that also limits the scope for radical multiple realizability. Few biologists seek causal dependences about wings, eyes, or other such broad functional categories. They will be interested in explaining dependences, regularities, or generalizations about insect wings or the cetacean or cephalopod eye. The narrower the functional category, the easier it is to honor the demands of MDBr.

One upshot for demands such as MDBr is obvious. MDBr can't tolerate much convergent evolution: convergent evolution produces functional kinds that are instantiated by radically different realizations, with quite different details of implementation that are so different from one another that they may not share even an abstract mechanism in common. When an environment begins to channel a variety of preexisting mechanisms into a narrow range of functions, it may result in unmanageable multiple realizability, the barrier to mechanistic and reductionist lower-level explanations that antireductionists have long insisted upon.

Conclusion: Mechanism, Causation, Physicalism, and the Laws of Nature

Mechanists are fond of "boxology" in biology. "Boxology" labels the invocation of three kinds of boxes – black, gray, and transparent – to delineate and isolate parts of a system that are initially black (not understood causally), gray (somewhat clarified in their causal processes), and transparent (fully understood as mechanisms). Craver and Darden (2013, pp. 89–90) write:

> A superficial, phenomenal model … describes the behavior of the mechanism without describing how the mechanism works … Incomplete schemas [that reveal the mechanism responsible for the phenomena] are best thought of as explanation sketches. They have black boxes for components for which not even a functional role is known … Sketches may also have grey boxes, for which a functional role has been conjectured … The goal in providing a complete description of a mechanism is to fill in black and grey boxes.

They go on to note that mechanism's methodology is "a version of the familiar childhood game of iteratively asking 'Why?', except in this case we ask, 'And how does that work?'" (p. 90). The process by which successive boxes are rendered transparent is by identifying causal difference makers inside them, components whose changes are difference makers for changes in the phenomena, the behaviors the mechanisms engage in. And these differences makers will be physical.

Bechtel (2008) is typical of mechanists when he writes: "In most biological disciplines, both the phenomena themselves and the operations proposed to explain them can be *adequately characterized* as involving physical transformations of material substances" (p. 23, emphasis added). Of course, as Craver and Darden (2013) write, "[I]t's not part of our [the mechanists'] view that all explanations must bottom out in some privileged set of fundamental entities and activities (such as elementary particles and strings). Biological explanations rarely need to descend to the depth of quantum physics. As currently understood, most biological mechanisms are otherwise *insensitive to differences in particular details of the components at such very small size scales*" (p. 90, emphasis added).

Mechanists' appeals to physical movements and physical processes reflect a fundamental philosophical commitment, one they share with reductionists. In the reductionist's original picture, the explanatory power of an explanation was conveyed by its appeal to the laws of nature. At the outset, reductionists were loath to invoke appeals to causation as what conferred explanatory power owing to a positivist desire to avoid the metaphysical baggage thought to attend the word "cause." Instead, following Hume, they treated causal claims as implicitly underwritten by laws and hoped that therefore causal explanation could be cashed in for nomological explanation. The role of laws would confer empirical content on explanations owing to their symmetrical role in prediction that implicitly tested scientific explanations. The reductionist's ordering of sciences from more basic to less basic, from more fundamental and general to higher level and domain specific, was grounded on the derivation of laws. It was their way of honoring the implications of a metaphysical commitment to physicalism as the motivation for methodological reductionism without metaphysical baggage.

Many mechanists recognize that in their analysis mechanisms take the place of laws as the drivers of explanation in the life sciences. Since there are no laws in these disciplines, there is nothing in the life sciences for classical reductionists to demand (further) explanation of. But if there are no laws, then something else has to do the work laws were supposed to do: explain, test, and predict. Since positivism long ago went into eclipse and there is no longer embarrassment among philosophers regarding appeals to causation, mechanists are happy to accord mechanisms causal explanatory force and test them via the predictions they allow us to make.

In fact, if mechanists can vindicate the explanatory power of mechanisms, they will have helped us to understand why biologists often treat the *ceteris paribus* regularities of their discipline as explanatory where they work, even while accepting that they often fail. Mechanisms explain their successes and enable us to diagnose their failures. It has been puzzling in the philosophy of science as to why and how *ceteris paribus* regularities secure their counterfactual supporting powers or whether indeed, owing to the ways in which they break down, they explain without having such powers. Insofar as mechanisms explain both when *ceteris paribus* regularities work and when they break down, they may solve this mystery, but only if the mechanisms carry the sort of counterfactual supporting force that laws do.

Mechanisms are composed of difference makers, subsystems whose behaviors are each individually causally necessary components of one or more sets of components whose joint behavior is sufficient for the occurrence of the phenomenon in which the mechanism consists. Following Woodward (2002, 2005), many mechanists base the explanatory power of mechanisms on this fact about them, and they appeal to Woodward's interventionist strategy both to explicate and to empirically test claims about causal relations between components and the mechanisms that contain them. Of course, as noted earlier in this section, Woodward's account of causation as a relation revealed by interventions on one variable that bring about changes in another variable is, so to speak, made for mechanism.

A local "invariance" between two factors, X and Y, is causal and therefore explanatory roughly if "interventions" – changes to the state of X that are followed by changes to the state of Y – and two other conditions obtain: The intervention changes Y only via the change it makes on X and not via its impact on any other factor causally necessary or sufficient for changes in Y; and the change to the state of X is not correlated with (changes to) any other factors that change Y. Such local invariances are explanatory because they are stable – that is, they hold even in the presence of interventions on other factors that might influence the occurrence of Y (Woodward, 1997, pp. 530–532).

Woodward's approach has many virtues as an account of which local and *ceteris paribus* regularities or invariances are in fact held to be explanatory in the life sciences. It certainly comports well with the mechanists' insistence that mechanisms explain, as well as with MDBr, the demand that more causally relevant mechanical details are always better, for the criterion of causal relevance that mechanists employ is Woodward's scheme of invariance under interventions (cf. Craver, 2007). It's obvious how to apply the Woodward recipe for causal difference makers in the multilayered diagram shown in Figure 1. Each of the X's bears an invariant relation to the phenomenon and no non-X bears such a relation, as experiments on hypothesized interveners will confirm. But of course the fact that X's are packaged together into a mechanism that explains S's ψ-ing is not a nomological fact – it is a local regularity, one that may break down, has already broken down elsewhere, has been prevented from emerging elsewhere, or ceases hereafter to be invariant owing to rare, novel, or unknown interventions. Thus, both local regularities that mechanisms explain, and regularities about how mechanisms bring about phenomena, will have to contain *ceteris paribus* clauses. And in the life sciences we know why this will always be the case. Both sorts of local invariance will have been put in place by the process – the mechanism – of natural selection operating on local conditions. Just as the functions that components perform and the adaptations that whole systems evince are the products of natural selection, so too will their breakdowns, malfunctions, exceptions, unraveling, and failures be the results of the changes – environmental, competitive, and otherwise – on which natural selection relentlessly works to build "better mousetraps" – improved adaptations. The full explanation of any local invariance will therefore have to include the operation of natural selection keeping it in place (Rosenberg, 2012). Natural selection's ubiquity will, of course, also be the source of its invisibility.

In the present context, Woodward's analysis may have one vice. As Woodward freely admits, the invariance/intervention approach to explanatory causal difference makers helps itself to a healthy dose of causal notions. It does not purport to be a "reductive" analysis of causation, one that cashes in the causal relations for some set of noncausal properties that jointly constitute causation. As a methodological prescription for reliably identifying complex causal relations or ones that are difficult to isolate or detect by starting with simpler or more obvious causal relations, this non-reductive feature of Woodward's account is no defect. However, it may not be enough of an account of causation to do everything the mechanists want their claims to do. Several mechanists have addressed the matter of what the causal relation that obtains within mechanisms and between components and the phenomena they produce consists in. Almost the full range of alternative analyses of the causal relation – the "cement of the universe" in Hume's words – are formally compatible with mechanism. But the attractions of a "quantity conserved in transmission" analysis are obvious. This is an approach that takes causation to consist in the transfer of a conserved quantity – momentum, for example, transmitted by physical contact or physical fields (Dowe, 1992). Mechanisms must "bottom out" into relations between physical objects and systems that preserve such quantities as they are transferred. There is only one way in which this can happen: the operation of physical law. In fact, many mechanisms that life scientists actually uncover rely on well-established laws of fundamental physics and chemistry. Consider the role of the laws of organic chemistry and biochemistry: every link in the TCA cycle in cellular metabolism is etched in physical law, as is the mechanism of respiration, or photosynthesis, all the way down to the quantum mechanics of individual photons. Consider the role of biochemistry in molecular biology: In the synthesis of nucleic acids, the entire mechanism of semiconservative double-stranded DNA synthesis that preserves the genetic information during gene duplication is fixed by the operation of the laws of organic chemistry. Another example, due to Weber (2005), is the role played by a fundamental physical regularity – the Nernst equation – in describing the electrical potential across a membrane at equilibrium in the mechanism of action potentials in neural signaling. In fact, Weber (2005) has argued, physical laws underlie the causal relations that obtain in mechanisms and secure the fundamental explanatory power of all mechanisms with regard to the phenomena they constitute. If this is right, then the mechanists' commitment to MDBr as a methodological prescription will in the end vindicate the physicalist hope that originally drove the reductionist research program at its origins 70 years ago. It will vindicate the explanatory fundamentality in all of the life sciences of physical law. By the same token, physicalism needs mechanism in order to extend the reach of these fundamental physical laws into the molecular biological, cell physiological, genetic, developmental, evolutionary, ecological, and all other functional domains of the life sciences.

Bibliography

Batterman, R., & Rice, C. (2014). Minimal model explanations. *Philosophy of Science*, **81**, 349–376.

Bechtel, W. (2008). *Mental Mechanisms*. New York: Psychology Press.

Bechtel, W. (2009). Looking down, around, and up: Mechanistic explanation in psychology. *Philosophical Psychology*, **22**, 543–564.

Bechtel, W., & Abrahamsen, A. (2005). Explanation: A mechanistic alternative. *Studies in History and Philosophy of the Biological and Biomedical Sciences*, **36**, 421–441.

Bechtel, W., & Levy, A. (2013). Abstraction and the organization of mechanisms. *Philosophy of Science*, **80**, 241–261.

Brandon, R. (1990). *Adaptation and Environment*. Princeton: Princeton University Press.

Cartwright, N. (1983). *How the Laws of Physics Lie*. Oxford: Oxford University Press.

Chirimuuta, M. (2014). Minimal models and canonical neural computations: The distinctness of computational explanation in neuroscience. *Synthese*, **191**, 127–153.

Craver, C. (2007). *Explaining the Brain*. New York: Oxford University Press.

Craver, C. (2015). Levels. In T. Metzinger & J. M. Windt, eds., *Open Mind*. Frankfurt am Main: MIND Group. Chapter 8.

Craver, C., & Bechtel, W. (2007). Top-down causation without top-down causes. *Biology and Philosophy*, **22**, 547–563.

Craver, C., & Darden, D. (2013). *In Search of Mechanisms*. Chicago: University of Chicago Press.

Craver, C., & Kaplan, D. M. (2011). The explanatory force of dynamical and mathematical models in neuroscience: A mechanistic perspective. *Philosophy of Science*, **78**, 601–627.

Craver, C., & Kaplan, D. (2020). Are more details better? On the norms of completeness for mechanistic explanations. *British Journal for the Philosophy of Science*, **71**, 287–319.

Crick, F. (1966). *Of Molecules and Men*. Seattle: University of Washington Press.

Cummins, R. (1975). Functional analysis. *Journal of Philosophy*, **72**, 741–764.

Darden, L. (2006). *Reasoning in Biological Discoveries*. Cambridge: Cambridge University Press.

Dawkins, R. (1982). *The Extended Phenotype*. San Francisco, CA: Freeman.

Dowe, P. (1992). Wesley Salmon's process theory of causality and the conserved quantity theory. *Philosophy of Science*, **59**, 195–216.

Dray, W. (1957). *Law and Explanation in History*. Oxford: Oxford University Press.

Dupres, J. (1993). *The Disorder of Things: Metaphysical Foundations of the Disunity of Science*. Cambridge, MA: Harvard University Press.

Feyerabend, P. (1964). *Reduction, Empiricism and Laws. Minnesota Studies in the Philosophy of Science, Vol. III*. Minneapolis: University of Minnesota Press.

Fodor, J. (1974). Special sciences (or: the disunity of science as a working hypothesis). *Synthese*, **28**, 97–115.

Fodor, J. (1975). *The Language of Thought*. New York: Crowell.

Garson, J. (2013). The functional sense of mechanism. *Philosophy of Science*, **80**, 317–333.

Gould, S. J., & Lewontin, R. (1979). The spandrels of St. Marco and the Panglossian paradigm. *Proceedings of the Royal Society of London*, **B205**, 581–598.

Havsted, J. (2011). Problems for natural selection as a mechanism. *Philosophy of Science*, **78**, 512–523.

Hull, D. (1974). *The Philosophy of Biological Science*. Englewood Cliffs, NJ: Prentice Hall.

Hull, D. (1989). *Science as a Process*. Chicago: University of Chicago Press.

Kaplan, D. M. (2011). Explanation and description in computational neuroscience. *Synthese*, **183**, 339–373.

Kellert, S. H., Longino, H., & Waters, C. K. (2006). *Scientific Pluralism*. Minneapolis: University of Minnesota Press.

Keys, D. N., Lewis, D. L., Selegue, J. E., et al. (1999). Recruitment of a *hedgehog* regulatory circuit in butterfly eyespot evolution. *Science*, **283**, 532–534.

Kim, J. (2005). *Physicalism, or Something Near Enough*. Princeton: Princeton University Press.

Kitcher, P. (1978). Theories, theorists and theoretical change. *Philosophical Review*, **84**, 519–547.

Kitcher, P. (1984). 1953 and all that: A tale of two sciences. *Philosophical Review*, **93**, 335–373.

Kitcher, P. (1995). *The Advancement of Science*. Oxford: Oxford University Press.

Kitcher, P. (1999). The hegemony of molecular biology. *Biology and Philosophy*, **14**, 195–210.

Kuhn, T. (1961). *The Structure of Scientific Revolutions*. Chicago: University of Chicago Press.

Lange, M. (1995). Are there natural laws concerning particular species? *Journal of Philosophy*, **112**, 430–451.

Lawrence, P. (1992). *The Making of a Fly*. Oxford: Blackwell's.

Levy, A. (2013). What was Hodgkin and Huxley's achievement? *British Journal for the Philosophy of Science*, **65**, 469–492.

Lewontin, R. (1978). Adaptation. *Scientific American*, **239**, 212–228.

Longino, H. (2013). *Studying Human Behavior: How Scientists Investigate Aggression and Sexuality*. Chicago: University of Chicago Press.

Longino, H. (2014). Pluralism, social action, and the causal space of human behavior. *Metascience*, **23**, 443–459.

Machamer, P., Darden, L., & Craver, C. F. (2000). Thinking about mechanisms. *Philosophy of Science*, **67**, 1–25.

Mayr, E. (1981). *The Growth of Biological Thought*. Cambridge, MA: Harvard University Press.

Mitchell, S. (2000). Dimensions of scientific law. *Philosophy of Science*, **18**, 295–315.

Monod, J. (1971). *Chance and Necessity: An Essay on the Natural Philosophy of Modern Biology*. New York: Alfred A. Knopf.

Nagel, E. (1961). *The Structure of Science*. New York: Harcourt, Brace and World.

Neander, K., & Rosenberg, A. (2009). Are homologies (selected effect or causal role) function free? *Philosophy of Science*, **76**, 307–334.

Nickles, T. (1973). Two concepts of intertheoretical reduction. *Journal of Philosophy*, **70**, 181–201.

Nijhout, F. (1994). Genes on the wing. *Science*, **265**, 44–45.

Oppenheim, P., & Putnam, H. (1958). *The Unity of Science as a Working Hypothesis. Minnesota Studies in the Philosophy of Science, Vol. II*. Minneapolis: Minnesota University Press.

Perutz, M. (1962). *Proteins and Nucleic Acids: Structure and Function*. Amsterdam and London: Elsevier.

Piccinini, G., & Craver, C. (2011). Integrating psychology and neuroscience: Functional analysis as mechanism sketches. *Synthese*, **183**, 283–311.

Potochnik, A. (2017). *Idealization and the Aims of Science*. Chicago: University of Chicago Press.

Quine, W. V. O. (1960). *Word and Object*. Cambridge, MA: Harvard University Press.

Rosenberg, A. (2012). Why do spatiotemporally restricted regularities explain in the social sciences? *British Journal for Philosophy of Science*, **63**, 1–26.

Salmon, W. (1984). *Scientific Explanation and the Causal Structure of the World*. Princeton: Princeton University Press.

Shaffner, K. (1967). Approaches to reductionism. *Philosophy of Science*, **34**, 137–147.

Simon, H. (1996). *The Sciences of the Artificial*, 3rd ed. Cambridge, MA: MIT Press.

Skipper Jr., R. A., & Millstein, R. L. (2005). Thinking about evolutionary mechanisms: Natural selection. *Studies in the History and Philosophy of Biological and Biomedical Sciences*, **36**, 327–347.

Sober, E. (1993). *Philosophy of Biology*. Boulder, CO: Westview Press.

Sober, E. (1999). The multiple realizability argument against reductionism. *Philosophy of Science*, **66**, 542–564.

Strevens, M. (2008). *Depth*. Cambridge, MA: Harvard University Press.

Tayberry, J. (2014). *Beyond Versus: The Struggle to Understand the Interaction of Nature and Nurture*. Cambridge, MA: MIT Press.

van Fraassen, B. (1985). *The Scientific Image*. New York: Oxford University Press.

Watson, J. B. (1965). *Molecular Biology of the Gene*. London: Pearson.

Weber, M. (2005). *Philosophy of Experimental Biology*. Cambridge: Cambridge University Press.

Wilson, E. O. (1999). *Consilience: The Unity of Knowledge*. New York: Vintage.

Wolpert, L. (1998). *Principles of Development*. Oxford: Oxford University Press.

Woodward, J. (1997). Explanation, invariance, and intervention. *Philosophy of Science*, **64**(Suppl. Pt. II), S26–S41.

Woodward, J. (2002). What is a mechanism? A counterfactual account. *Philosophy of Science*, **69**, S366–S377.

Woodward, J. (2005). *Making Things Happen: A Theory of Causal Explanation*. New York: Oxford University Press.

Wright, L. (1976). *Teleological Explanation*. Berkeley: University of California Press.

Cambridge Elements ☰

Philosophy of Biology

Grant Ramsey

KU Leuven

Grant Ramsey is a BOFZAP research professor at the Institute of Philosophy, KU Leuven, Belgium. His work centers on philosophical problems at the foundation of evolutionary biology. He has been awarded the Popper Prize twice for his work in this area. He also publishes in the philosophy of animal behavior, human nature and the moral emotions. He runs the Ramsey Lab (theramseylab.org), a highly collaborative research group focused on issues in the philosophy of the life sciences.

Michael Ruse

Florida State University

Michael Ruse is the Lucyle T. Werkmeister Professor of Philosophy and the Director of the Program in the History and Philosophy of Science at Florida State University. He is Professor Emeritus at the University of Guelph, in Ontario, Canada. He is a former Guggenheim fellow and Gifford lecturer. He is the author or editor of over sixty books, most recently *Darwinism as Religion: What Literature Tells Us about Evolution*; *On Purpose*; *The Problem of War: Darwinism, Christianity, and their Battle to Understand Human Conflict*; and *A Meaning to Life*.

About the Series

This Cambridge Elements series provides concise and structured introductions to all of the central topics in the philosophy of biology. Contributors to the series are cutting-edge researchers who offer balanced, comprehensive coverage of multiple perspectives, while also developing new ideas and arguments from a unique viewpoint.

Cambridge Elements ≡

Philosophy of Biology